Automating the Welding Process

Successful Implementation of Automated Welding Systems

James M. Berge

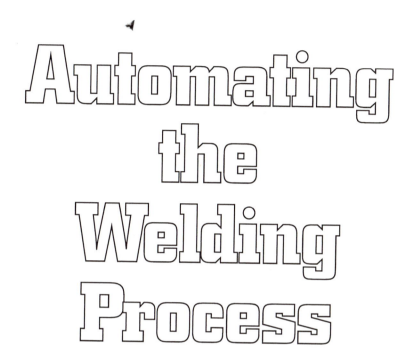

Automating the Welding Process

Successful Implementation of Automated Welding Systems

Industrial Press Inc
New York

Library of Congress Cataloging-in-Publication Data

Berge, James M., 1962–
  Automating the welding process : successful
implementation of automated welding systems / James M.
Berge.—1st ed.
    208 p.   15.6 × 23.5 cm.
    Includes index.
    ISBN 0-8311-3051-2
    1. Welding—Automation.   2. Robots, Industrial.
TS227.2.B49   1993
671.5′2—dc20   93-27580
                                                    CIP

Copyright © 1994 by Industrial Press Inc., New York, New York.

INDUSTRIAL PRESS INC.
200 Madison Avenue
New York, New York 10016-4078

First Edition

Automating the Welding Process

First Printing

10   9   8   7   6   5   4   3   2   1

Contents

Preface

Welding technology has been changing more rapidly than perhaps any other manufacturing process. It seems that there are new breakthroughs daily that promise higher quality and optimum productivity. One could easily get lost in the multitude of power supplies, welding torches, welding wires, and gases now available. It is difficult to know which ideas are even worth pursuing. One solution that almost every welding manufacturer should consider, however, is welding automation.

The proper installation of automated welding machinery in your facility could increase the output of your welding department by two to four times. Such equipment can increase the quality and consistency of your welds, while providing a safer and cleaner atmosphere for your welding operators. Welding bottlenecks could disappear, eliminating expensive and burdensome work in process, while the flexibility provided by robots and other types of equipment can help you to decrease your lot sizes to as small as one.

If all of your welding is still done manually, and if you have bid against a company who can provide the speed, quality, and efficiency that a robot provides, then you already know what you are up against. The most successful manufacturers are utilizing robots and other forms of welding automation to become as competitive as they can be. Their customers prefer to purchase components with consistent, high-quality welds, at lower costs. Companies with automated welding are not so prone to such factors as the increasing lack of qualified welders for hire or the drop in productivity when the weather becomes too hot. You have most likely automated many other production processes in your plant. It is now time to tackle the welding.

The benefits of automated welding are clear, but how can a company like yours be assured that its efforts will prosper? There are two important ingredients in the equation for success. First, a thorough understanding of automated welding machinery, techniques, and philosophy is necessary. This book will help you comprehend these variables and will describe the issues that must be addressed in order to maximize the profitability of your system and to optimize its performance. By becoming more knowledgeable about the reasons for automating, the various types of automation available, and how to calculate returns on robotic and automated machinery investments, you will learn just what you have to gain. You will then be in a better position to discuss automation with vendors, to recognize a good candidate for automation in your factory, and to know exactly how to begin the task of automating.

The second necessary ingredient for success is a result of the first, and that is the elimination of the fear and doubt that develops from a lack of practical knowledge. By understanding the welding process, and how to bring the various aspects of it under machine control, you will be one step closer to achieving all of the benefits that derive from a correctly developed automation program. By understanding the philosophy of automation, you will be able to avoid many of the mistakes and setbacks that other companies have made. If you have heard stories of welding robots that have failed, let me assure you that most of those failures could have been avoided with the application of the principles in this book.

World class manufacturers around the world have achieved great success with robots and automated machines. Since your competition is constantly becoming more global, automation of your welding operations is not only feasible but also becoming more necessary daily. The day will come when automation becomes essential for survival. The fact that you are reading this book proves that you are not content to maintain the status quo or merely to survive. You want to excel, and welding automation is an important rung in that ladder of success.

1. Why Automate?

The most relevant answer to "Why automate?" is this: "Because your competitors are." These competitors are not only here in the United States—YOUR COMPETITION IS NOW GLOBAL! The security of limited competition is gone, and you must learn to cope with other countries' lower cost of labor, more elaborate government incentives for manufacturers, and a level of automation that exceeds that in the United States. Chances are you have long ago automated your machining processes. Now it is time for you to tackle your welding.

Introduction to Automated Welding

For our purposes, automated welding is any welding process where certain motions of the welding arc are manipulated by a machine, rather than a person. Automated welding refers to equipment that will save you money and increase your profits. Automated welding does not necessarily mean robots—although I will be discussing robots at length in this book.

To assume immediately that any automation of your welding operations requires a robot could be a very costly mistake, especially in those instances where some other type of automation could outperform a robot. My goal is to bring you to the point where you can at least recognize a candidate for automated welding when you see one in your shop and to give you a general understanding of the types of automation available for welding, the returns you can expect from such an investment, and how to go about implementing this type of equipment. I will walk you through the entire process, from the time

1

you decide that you want to save money, increase your quality, and become more competitive, to the actual startup of a piece of equipment on your shop floor.

Welding is unique!

Welding is different from most other fabrication and CNC processes, such as machining. The variables are numerous, and the combinations of variables are infinite. Even the position of the molten weld puddle relative to gravity is critical to weld quality. The importance of all of these variables is evident when you compare experienced welders to rookies. I will attempt to reduce the welding process—as much as is possible and feasible—to quantitative facts; to stress the science over the art. However, there is no substitute for the experienced welder who knows the "art" as if it were second nature. Therefore, the welders must be involved as much as possible in the process of automating your welding.

Use this book as a textbook. You can use the table of contents and the index to look up particular topics, but I would encourage you first to read it from beginning to end, like a novel. Pay attention to the illustrations, and do not continue until you understand the concepts being discussed and how they apply to your specific applications. By the time you have finished digesting this document, you will be well equipped to attack your welding operations and finally beat them into submission. I hope that, in the end, you will realize that automated welding is one of the most lucrative investments you can make.

A Little History

About 4400 robots per year were installed in the United States in 1990 and 1991, according to the Robotic Industries Association in Ann Arbor, MI. Of these nearly 9000 robotic installations, more than 3500 were used for welding (see Figure 1-1). Many of these robotic welding installations are probably in your competitors' factories. This is reason enough to automate, but what you also need to consider are the plethora of welding robots installed in Japan and in Europe, not to

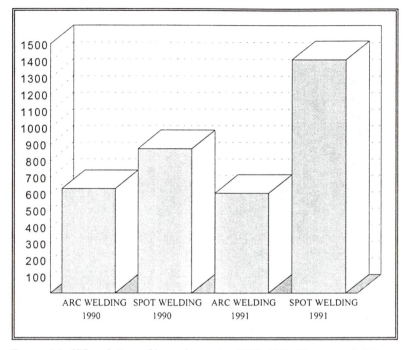

Figure 1-1. U.S. robotic welding shipments, 1990–1991 (Courtesy of Robotic Industries Association)

mention the other more abundant types of automated welding machines. The days of secure stateside competition are over. We are now competing for our lives in a global market, and many others in that market are far in the lead where welding automation is concerned.

In general, welding automation must not be viewed as a way to decrease your workforce in order to increase your bottom line but instead as a way to become more competitive. History shows that, most often, companies that implement welding robots end up growing and having to hire more welder/robot operators. Why? Obviously, significant productivity improvements are realized, but, at the same time, quality typically improves dramatically. Labor savings— usually the primary (and often, erroneously, the only) criterion used in the payback calculations—can be phenomenal. Significant indirect labor savings are possible, due to improved material flow and the elimination of welding bottlenecks. The possibility of providing higher quality, lower priced weldments in small lot sizes (as small as one) allows you to fill orders more effectively, meaning that you provide customer service that is superior to that of your competition. With robotics, the programmability of the equipment provides a lot of flexibility in running several part numbers at once. This is most beneficial where you are welding subassemblies that are assembled into a final product.

For instance, if your end product is riding mowers, you may need to weld several subassemblies first. The axles, the mower deck, the frame of the vehicle, and the engine mount may all be welded prior to a final welding process where all of these components come together in a common welding fixture. Robot users may have the ability to weld one of each of these subassemblies with no changeover and then weld the subassemblies into a complete riding lawnmower frame. In this way, you will be providing true just-in-time delivery to the final assembly and paint area.

The reasons stated above are just a few of the reasons to automate your welding operations. There are more, and most of these can and should be quantified in your equipment justification. I am convinced that if you effectively implement welding automation, you will find that the true savings are much more than you originally expected. The Japanese and Europeans have discovered how to make their welding operations more efficient and profitable. Now American manufacturers are, too. To keep their edge, they have recognized the value of automating what is probably the last process in their plants that has not yet been automated. To keep your edge, you need some knowledge of welding and automation. It is a new way of doing business, a new way to become the best at what you do, and a new way to invest in the successful future of your company.

Personnel Issues

One of the problems that other parts of the world have been feeling for some time, and one that is becoming more prevalent here in the United States, is a lack of qualified welders to hire. In some places, this is not yet an issue, but many are predicting the lack of laborers in all skilled trades, not just welding. Welding automation gives a company the ability to limit the number of highly skilled workers, if such people are not available. As long as there is a qualified welding technician around, a nonskilled laborer can load and unload a welding machine. I am not suggesting that this is the "ideal" situation. This is simply one reason why robots are becoming more necessary.

Please do not make the mistake of thinking that just anyone can effectively program a welding robot. It takes about a week to learn the programming language of almost any robot available today. It is easy for welders, but it is not at all easy, or even possible, to teach a nonwelder all of the intricacies and variables involved in welding in such a short time. Since the operator will be imparting his or her knowledge of welding to the machine in order for it to produce quality parts, that operator must be intimately familiar with the welding process. This is why we want to address the issue of just who should be involved in such a program to automate. By putting the correct team together, you can work wonders.

Increased Productivity

The chief advantage of robotic welding is the ability to complete more weld cycles in a given time. A statistic that is very typical in the industry is that a welding robot can do about three to four times the work of a single welder, but this can vary from two to six times or more. This depends on several factors, including the following:

• *Ratio of arc time to total cycle time.* If the total time that the arc is on and the machine is actually welding is a high percentage of the total cycle time, then returns on your investment may be less than average. This is due to the fact that a machine, in general, will not actually weld much faster than a person can. The machine's greatest asset is the ability to do tasks in parallel with the operator that would otherwise be done in series by the operator alone. On the other hand, if the majority of the total cycle time is air cut

moves—the movement of the welding torch from weld to weld—
then the increase in productivity from automation is more
dramatic.

- *Weld sizes required and relative length of welds.* To automate
  extremely heavy duty welds or welds with multiple passes, the
  investment in automated welding equipment tends to increase.
  More sophisticated technologies are needed in order to handle the
  variability that is inherent in a multiple pass weld. Multiple
  passes also increase the ratio of arc-on time to total cycle time,
  which affects the factor discussed just above. As far as lengths of
  welds are concerned, the same concept applies: the longer the
  welds, the higher the arc-on time, and the less you can take
  advantage of a robot's 3-m/sec air cut speed.

- *Amount of part manipulation necessary during the weld cycle.* A
  healthy percentage of automated welding systems are engineered
  with some means to manipulate the weldment during welding. You
  realize that the only time that machine will earn you money is
  when it is actually laying weld metal down on the parts. All other
  activities are extraneous to the welding itself and are not actually
  adding value to the component. Actual welding usually cannot be
  accomplished during part manipulation, so minimizing this
  activity increases the value of your investment.

- *Whether operator intervention is required during the weld cycle.* In
  some instances, certain welds may be located inside a cavity or
  channel that is then capped and welded. These inside welds must
  first be completed, then the additional components are welded on.
  If the volume of welding to be done in such a manner is
  significant, the machine may be instructed to wait, at a certain
  point in the weld cycle, for the operator to load additional parts
  and restart the system. The time it takes the operator must usually
  be added to the total machine cycle.

- *What degree of automation is already present in the process.* If
  electric, foot-operated trunnions are now being used to rotate
  assemblies during welding, then that represents a certain amount
  of automation already being utilized. This will change the

estimates in your productivity increases as compared to automating a process that utilizes absolutely no such mechanization.

Higher duty cycle

The first main increase comes from the high air cut speeds possible with automated processes (the speed at which the welding torch moves from weld to weld). Robots have air cut speeds of 2 to 3 meters (6.5 to 9.8 feet) per second and more. Where the welder must raise his or her helmet and look where he or she is going next, the robot can move to the next weld in less than a second. A good welder has the arc on about 30% of the time, compared to a typical robot, with a duty cycle of about 85%. It is easy to determine that the high air cut speeds have a significant impact on the total weld cycle (Figure 1-2). Some manual welders have discovered the benefits of welding hoods that darken automatically, increasing manual duty

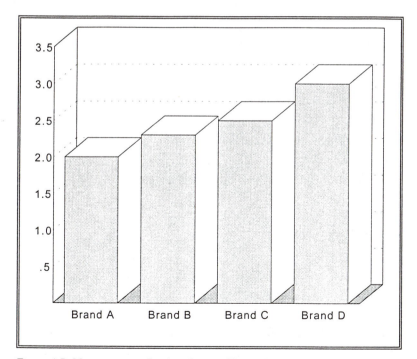

Figure 1-2. Maximum speeds of modern welding robots (m/sec)

cycles. These are excellent productivity boosters and should be seriously considered. However, I have seen these helmets used as an alternative to welding automation, while automation provides significant gains that these helmets just cannot address.

Internal load/unload

The next major advantage is internal loading and unloading of parts. The operator can unload the finished part and reload the fixture while the machine is welding at another station. The operator can also do any necessary tack welding at this point. This puts the load/unload time and tacking time completely internal to the welding time. Figure 1-3 shows a machine with an indexing table. Station 1 is in the machine being welded, while Station 3 is being unloaded and reloaded by the operator. The loading time is free, since it is internal to the weld cycle. In the same way, Figure 1-4 shows a robotic welding cell with two stations. Station 1 can be unloaded and reloaded while the robot welds at Station 2. This can give the operator ample time not only to reload the fixture, but also to replenish parts, do any necessary tack welding, or even complete welds that the robot could not reach.

On larger, more complex weldments, the tacking and fitting time can easily reach 25–50% of the total cycle time. With manual welding, these tasks are done in series, lengthening the cycle time significantly. With a robotic system, the tacking is done parallel to the welding. This alone can double the output of a robot when compared to manual welding, even if the robot welds only as efficiently as a human being.

Part manipulation

Most automatic positioners can rotate loads a full 360 degrees in a matter of seconds. Manufacturers that currently use hoists and cranes to manipulate heavy parts for manual welding can see dramatic reductions in cycle time by implementing even a simple manually operated positioner. However, combine a programmable, fully integrated servo operated positioner with a robotic welding cell, and the productivity gain from this single factor can be astounding.

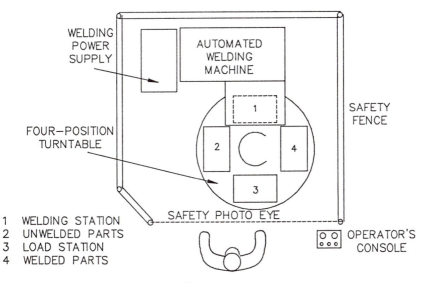

Figure 1-3. Four-station indexing welding machine.

One manufacturer of very heavy steel components had a manual welding time for a certain weldment of 27 hours. It was proven that a robotic welding system could reduce the cycle time to about 4 hours. One reason for the dramatic improvement was that extensive part manipulation was required during the weld cycle. Since the parts

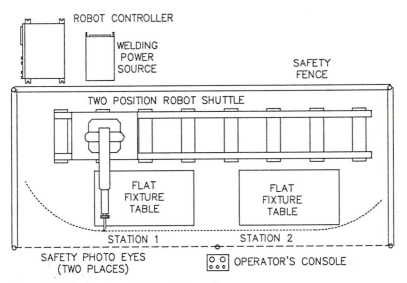

Figure 1-4. Two-station robotic welding cell.

weighed several thousand pounds, they had to be flipped several times with an overhead crane, a very time consuming process. Multiple pass welds on both sides of the part, as well as the steel alloy of the base metal, required up to 15 passes—no weaving allowed—while maintaining a particular interpass temperature. It was a very labor-intensive operation and one that the operators did not like at all (imagine welding on a steel plate, which has been heated to 500°F, while the ambient plant temperature is 100°F). The robotic welding system that was proposed included a positioner to turn the part over in a matter of seconds.

Decrease or eliminate changeover time

A fourth increase in productivity can be realized in installations where changeovers are frequent. By using universal fixturing and other concepts, changeover to another part number can be as simple as pushing a button. This can greatly reduce nonproductive labor time. It is possible for a flexible welding system to weld 20 different part numbers in a cell with only four workstations and have the robot never sit idle. The challenge here is for the other plant processes to keep the hungry beast fed with component parts.

The computer age has had a very positive effect on automated welding. Automatic identification of parts, mass storage of information such as welding programs, and artificial intelligence have all been integrated effectively into today's automated systems, and all of these capabilities can help to eliminate downtime and increase the productivity of your welding cell.

Increased welding speeds

Although automated welding processes can indeed deposit weld metal faster than a human being, it is not always possible. If you have talented welders, they will most likely be able to weld about as fast as a robot can. However, significant gains in welding speed can be achieved, depending upon the particular application. For example, where a person must slow down in order to weld around a complex contour, a machine may weld more quickly.

There is usually an ideal speed at which to deposit a certain size weld bead, and it may make no difference if a human or a machine is holding the torch. But if, for example, a 10% gain in weld speed is possible, then it can make a significant impact on the savings possible from such a welding system.

Increased weld quality

A major advantage to automating your welding, although some-times difficult to quantify, is the increased quality that results. Almost all automation has the ability to increase weld quality, since a machine can travel at precisely consistent speeds and can repeat that speed very accurately from weld to weld. Assigning value to that attribute can be challenging.

Some heavy equipment builders, after having to grind manual welds to a smooth contour before painting, find that the weld quality provided by their robots allows them to eliminate the grinding and send the welded parts directly to the paint booth. This is a number you can quantify, which relates directly to the consistency and quality of the weld bead.

Needless to say, if the enhanced aesthetic appearance of the welds exceeds that of your competition and allows you to obtain a new contract, then the advantage of higher weld quality is easily under-stood. At the same time, if your manual welds cause you to lose an order to a competitor with a robot, then the cost of NOT automating becomes painfully clear.

Operator Safety

OSHA is becoming stricter in their standards addressing the quality of the air that your workers are breathing. By making the welder an operator of a machine or robot system, you are removing him or her from a hot, smoky, unhealthy environment. There are proven methods for capturing the welding fumes close to the weldment, so that the overall air quality in the plant can improve also.

By installing flash screens and safety fences, an automated

system can be made truly safe. Let me emphasize that safety is a critical issue when dealing with automation of any sort. When an 800-pound robot makes sudden movements at 2 m/sec, it can easily injure someone in its vicinity. For that reason, most vendors are providing safety equipment as part of robot and automation packages. If they do not, make sure that they add it on.

My purpose is not to scare you about the safety of robots. Serious injuries are extremely rare from properly guarded robot cells, even though standards have not always existed. But since OSHA will be implementing even stricter guidelines, you need to realize that automated welding can effectively address certain safety issues.

Predictable Performance

A major advantage to automation of any kind is its predictability. It will provide consistently high quality parts every cycle. This, of course, depends on certain factors beyond the control of the machine itself, such as fitup of the component parts, repeatability in the manufacturing of these parts, and whether the machine has been set up correctly. This predictability gives you assurance, when bidding on new jobs, that you can actually achieve a certain production volume (which, with automation, will hopefully be higher than that of your competitors) and at a certain cost (hopefully lower than the competition).

Since a robot is a programmable machine, it will exactly repeat its tasks every time. The production rate of the robot should not drop off when the temperature in your plant reaches 95°F, or when deer season begins. A machine does not care whether it is welding on 500°F preheated steel (generally) or whether it is asked to work three shifts per day. And even after the third shift, it will continue to create the same quality as it did all day long.

This predictable performance is not a miracle cure. Many companies are continually in the firefighting mode, where they must struggle with every resource available just to meet daily, weekly, or monthly quotas. Automation will not mysteriously change the way your company operates, but it can relieve the scheduling headaches involved in guessing how long it should take a welder to weld a particular part.

Consistent Welding Variables

Consistent welding variables are closely related to consistent quality, as we have just discussed. However, many other advantages are seen by keeping those welding variables under machine control. For instance, it is a proven fact that people tend to overweld. This is not altogether bad, since overwelding may make up for certain inconsistencies arising from the manual welding process, but it can be incredibly expensive. (We will see why in Chapter 4.) The absolute consistency of these variables is the cause of the high weld quality available from automated applications.

Consistency is also critical when working with codes or other standards. Just as an operator can be approved and certified to a particular code, so too can a machine. In fact, it is often easier to certify an automated process than a manual process, due to the control of variables that a machine provides. This can provide unprecedented freedom and accuracy as you prepare proposals that require certified processes and that require a high degree of control and information, such as SPC figures on the welding parameters used.

Uniform, Predictable Quality

Consistent weld quality means higher consistency not only from part to part, but also from run to run. Assume you are a job shop, and you have a particular product that runs only once every two months. Once your robotic welding program is written, those parts will be welded identically every time that you run them. Identical welds mean consistent quality, which leads to happy customers. Happy customers provide repeat business, referrals, and business growth.

The quality provided by automation is often underrated. There are some great welders out there, but none of them can provide his or her absolute best quality eight hours a day, every day of the week, and do it consistently. No matter how hard it works, a machine has the ability to repeat itself precisely time and again. This means quality—and the highest weld quality available today comes from machines.

How many times have you heard the argument between first and second shifts about where the welding machines should be set? I would wager that each of the welders in your shop has a unique setting at which he or she prefers to run the welding machine. Consequently, a lack of quality in manual welding can come from a number of factors, including the following:

- Inconsistent welding variables can produce welds that are too cold or too hot; this causes overlapping or undercutting of the weld puddle and can severely limit penetration, causing weak weld joints.

- Inconsistent torch-to-work distance (stickout) directly affects the welding variables, since the amperage in a constant-voltage MIG welder is directly proportional to stickout. If severe, this can also lead to insufficient shielding gas coverage by putting the weld beyond the area of laminar gas flow. When flowing from the nozzle, the shielding gas flow is laminar, or smooth, and the weld is completely covered by shielding gas, preventing the atmosphere from touching the molten weld puddle. As distance from the nozzle increases, the flow becomes turbulent, which sucks contaminating air into the weld zone. Although the operator may think the gas coverage is there, the turbulence can let air in, which causes weld defects.

- Inconsistent torch angles will cause the weld to be positioned too high or too low in a fillet joint and can cause porosity from a compromised angle of the gas nozzle, can cause slag inclusions in flux-cored welding, and can severely limit penetration. Weld quality is especially dependent on torch angle when one considers the shielding gas being used. The typical procedure for welding with CO_2 is to use a drag angle, which reduces spatter and optimizes penetration. When using an argon mix for shielding gas, however, a push angle is usually recommended. By varying this angle during welding, manual welders can create welds with inconsistent penetration and with varying aesthetic quality.

* Inconsistent attitude of the weldment with respect to gravity is a very important variable, especially with certain types of welding wires. Flux-cored wires that are not designed for out-of-position welding are very sensitive to weld position and usually cannot tolerate a weld that is more than 5 to 10 degrees from horizontal without the molten puddle running out of the joint. Automated welding machines are designed to keep the weld in the proper orientation during welding.

There are many more such factors that affect weld quality. All of these parameters are critical for maintaining the integrity and quality of the weld, and all of them can be automatically controlled. These are among the many ways in which automation contributes to superior weld quality.

Predictable Material Usage

It is important to realize that a typical robotic welder can have up to three times the arc-on percentage as a human. This means that rather than changing contact tips every few days, you may need to do it every shift. A human does not care if there is a huge hole worn in the contact tip, but try that with a robot, and welds can be mispositioned due to the cast in the wire.

The same holds true for torch liners and wire feed drive rolls, as well as the wire feeder motor itself. Preventive maintenance (P.M.) can sometimes make or break a welding system. It is very important to implement such a program, no matter what level of automation you install, but such a P.M. program is very simple to establish. In short, with automated welding machines, the consumable use is predictable, and that helps you control costs more effectively.

Welding wire and shielding gas are the other consumables to consider. At the rate of 500 in./min, a robot will eat up a 60 pound spool of wire in a hurry. If the operator must change spools three times per shift, then the robot sits idle for up to 45 minutes each shift. The solution is to use bulk welding wire, available on spools or coils containing up to 1000 pounds of wire. Several commercially available bulk wire dispensers are available starting at around $500.00. Talk to your welding automation vendor about it, or contact your welding distributor for more information.

The following graph shows how deposition rate is affected by welding current. Higher amperage means higher deposition rates and, thus, higher productivity. However, at high amperages, the operator experiences great discomfort from the excessive heat. It is easy to see how productivity can suffer if the welder turns down the heat. At the same time, it is easy to see how quickly your welding wire will be consumed if you automate the process and weld at the highest practical amperage (Figure 1-5). At 300 amps, the welder may be depositing 12–13 lb/hr of weld metal. At 500 amps, however, deposition rates can be as high as 23 lb/hr. People do not like to weld at such high values, since the heat is excessive. A machine is not sensitive to this factor and may increase productivity favorably.

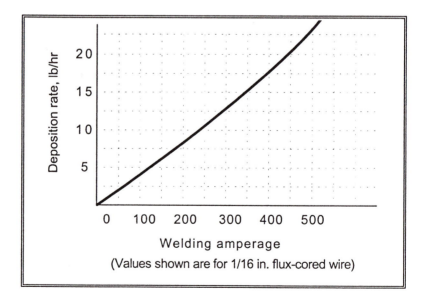

Figure 1-5. Deposition rate at different welding amperages.

Another advantage to buying welding wire on bulk spools is the cost. You can usually save a significant sum, since the manufacturer's labor cost is not as high per pound of wire. Because of its programmability and consistency, automation may also allow you to use a larger wire diameter than with manual welding. This also saves money because larger diameter wires are less expensive than smaller diameters, being drawn through fewer dies than the smaller wires are.

The same cost advantage is available through the use of bulk shielding gas, although the investment in a bulk gas system is considerably more than for a bulk wire dispenser. Typically, a single automated welding system cannot use gas fast enough to make a bulk tank operate efficiently. If not used quickly enough, certain bulk gases will release into the air to avoid pressure build-up, and this is like throwing your money to the wind. If you do not yet have a bulk tank, then consider it for your whole shop. The savings can be substantial. Your welding distributor can lead you down the correct path and provide information for a complete system, including all the indoor piping and drops to each individual welding station. Regardless of the method employed to dispense wire and gas, you will find that usage of these materials is very consistent and predictable with automation.

Limitations of Automated Welding

This discussion would not be complete without outlining the things that you should not expect from automated welding. Robotic welding got off to a rough start, since many saw it as a cure-all for their welding problems. Many companies spent money on (notice that I *did not* say "invested in") robots, only to find that they just did not work. Perhaps you are one of those who purchased a robot in the 1980s, could not make it work, and either decommissioned it or ended up using it in a job for which it was never intended. Those anecdotes abound, but remember that these things happened mostly during a time when robotic welding was in its infancy. It is typical for any new technology to struggle for some time before it is fully understood. The robotic welding industry has indeed had its ups and down during the last several years.

The first robots were hydraulic, and, compared to modern electric robots, were highly inaccurate. High-tolerance welding was not even attempted with these lumbering beasts, but they filled a very effective niche in the applications that suited them. They performed well enough, in fact, to cause the robotic industry to sit up and notice the potential. So next came the electric servo-controlled robots, which were quite accurate, much faster than their hydraulic ancestors, and seemingly much more appropriate for a wider variety of tasks. Yet vendors and end users alike often placed them in poor applications, or in jobs where the component part quality simply would not allow automation to succeed. Many fell by the wayside after causing a sour taste in the mouths of many end users.

Unfortunately, this scenario still sometimes takes place, although rarely, even in the maturing robotic welding market. This should underline the importance of getting involved with a vendor who has a good track record. Chapters 5 and 6 discuss the topic of evaluating vendors in greater detail. It is easy to remove the risk from the implementation of an automated welding cell. Thousands are utilizing such automation even as you read this book.

Inconsistent parts

The problems with these early installations—and even some today—were varied. One of the most common robot killers were the inconsistent parts people tried to feed the robot. A machine is blind (without some type of optional seam finding capability, which was not available years ago) and will do only what it is taught to do. It will repeat itself within a few thousandths of an inch every time. If the weld joint is not there, the weld metal will not be deposited in the joint. Inconsistent parts are no problem for people, with their built-in seam-finding capabilities and eye—hand coordination. A machine, however, wants to have the same thing presented to it every time. The good news is that by improving your component part quality for automated welding, you are also improving the overall quality of the final assembly.

Inconsistency also means gaps. If the weld joint is mispositioned, there are proven technologies (some expensive, some not) that can find it. If there are gaps, however, the solutions to compensating for them are few and far between and usually are very expensive. A man can easily slop the weld metal into a gap to fill it, but that decreases the joint strength, uses a lot more filler metal, and usually looks quite poor. By tightening up the gaps in your parts (which is something you should do even for manual welding), you are a step closer to automating. In case you are wondering, there are ways of finding weld joints if repeatability is a problem. This will be discussed in Chapter 2.

Figure 1-6 shows the effect of gaps on weld quality. Aside from making it more difficult to automate the welding, gaps will severely weaken a weld joint. Drawing A shows a fillet joint with tight fitup. Drawing B shows the same fillet, but with a gap that is half the thickness of the base metal. In $1/8$ in. sheet metal, this is a gap of only $1/16$ in. The area of penetration shown in View B (with a gap) is only 25% of that in View A (with no gap). It is indeed important to close up your gaps, whether welding manually or with automation.

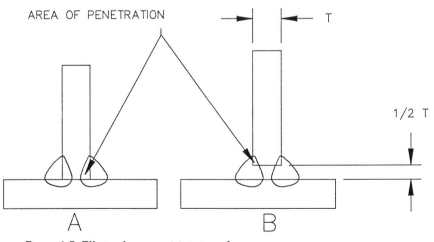

AREA OF PENETRATION

T

1/2 T

A   B

Figure 1-6. Effects of gaps on joint strength.

Part cleanliness

Part cleanliness can be an issue with automation. Many manufacturers today are shot blasting every part that gets welded by a machine. This may seem like a lot of trouble, but the rewards are significant. Again, manual welders can compensate for oil, dirt, mill scale, and rust by manipulating the puddle to wash out those extra impurities where it is needed. As discussed earlier, however, the very thing that provides high quality welds with automation—the consistent weld parameters—makes a machine unable to react to changing cleanliness conditions. I am not implying that you must shot blast your parts in order to weld them automatically. It is actually much more an issue on heavier plate that is subject to rust and mill scale and not as much an issue on tubular and sheet metal products.

It is important to remember that in making some of these recommended improvements in preparation for automated welding, you will not simply be making it easier for a machine to weld your parts. By shot blasting parts prior to welding, by improving the fitup between mating parts, and by manufacturing your components more accurately, you are improving the overall quality of your product. I suggest that many of these improvements should be made even if you do not plan to automate your welding. Such improvements can only add to the value and attractiveness of your product in a competitive environment.

Automation needs people

This next issue regarding limitations of automation is less a limitation than it is a problem with the philosophy of automation in general. I will later discuss who should be involved in such a project, but, for now, suffice it to say that from the very conceptual beginning of a project like this, the welders who will be running the system (yes, we want welders to learn to program and run the equipment) must be involved, their opinions must be respected, and they need to accept the whole process. They should not necessarily be given the ultimate say in whether to purchase the equipment, simply because they typically are not privy to such information as shop rates, overhead, financial justifications, amount of capital available, and the like (unless the management sees fit, as I recommend, that this information is made available). They must, however, be on the forefront of the project.

A certain manufacturer of heavy steel weldments has the right idea. A metamorphosis has been taking place during the last few years that is certainly not a new idea but is admittedly not that common in the United States. The factory floor workers were given authority.

After being split into three teams, according to the three basic types of products the company builds, the teams were given decision-making abilities and the authority to run their cell as they saw fit. A team manager, an engineer, and a buyer-planner work closely with the welders, assemblers, and machine operators to schedule production, part delivery, and purchases, decreasing the customer's lead time dramatically as compared to just a few years ago. Inventory has decreased significantly as the company heads closer and closer to its goal of being a just-in-time manufacturer.

The most recent purchase of a nearly half-million dollar robotic welding cell was entirely an effort of these team members themselves. The welders and machine operators were personally involved in presenting the project to the company's leaders, and had full knowledge of the cost, the development, the delivery, and the implementation of the equipment. As a result, they owned the project. The robotic welding system was theirs; they chose the vendor, they purchased the system, they were excited about the project, and they held the responsibility to make it work—and the robotic welding cell works today.

Sharing the work with the welder

If your weldments are large or complex or both, do not expect a piece of automation to do 100% of the welding. Often it is possible, and that is fine, but sometimes the difference between reaching 85% of the welds with a machine and reaching 100% of the welds can triple the cost of the system. By identifying where the lines meet, you can purchase the welding system with the best combination of value and productivity.

Figure 1-7 contains a comparison of how productivity can be improved by using a less expensive, easier-to-operate system to balance the work between the man and a robot. Where 100% of the welds are completed by the robot, the system is more expensive, since more flexible equipment is needed to manipulate the part properly. By simplifying the machinery, the end user reduces the capital investment, gives the operator less idle time, and decreases the cycle time, all simultaneously. By sharing the work, productivity is improved, the startup is faster and easier, and maintenance of the system is less troublesome and less costly.

100% of assembly welded by robot

Equipment cost	$250,000.00
Operator's prep time	30 min
Robot cycle time	60 min
Effective minimum cycle time	**60 min**

(Operator is idle for 30 min while robot welds.)

80% welded by robot, 20% manually

Equipment cost	$190,000.00
Operator's prep time	30 min
Operator's weld time	15 min
Robot cycle time	48 min
Effective minimum cycle time	**48 min**

(Operator is idle for only 3 min.)

Cycle time is reduced by 20% while equipment cost is reduced by $60,000.00. Lower equipment cost is due to purchase of a less flexible part positioner.

Figure 1-7. Comparison between 100% of welding by robot and 80% by robot, 20% manually.

Summary

This chapter has been devoted to discussing why you should automate your welding. I hope that the benefits are clear. There are more benefits that we will be discussing in detail, but we are talking about an investment that can have a very good return for you. You can increase your quality, increase your productivity, respond to customers with greater flexibility, and provide a safer atmosphere for your welders, all while saving lots of money. Sound too good to be true? Tell that to your competitors and the thousands of other companies who are now enjoying the fruits of their automation installations.

A. Introduction
   1. Welding automation does not necessarily mean robots.
   2. The operators must be involved!
B. A little history
   1. Welding automation expands your business.
   2. We can learn from the mistakes of the past and avoid the growing pains of the robotic welding industry.
C. Personnel issues
   1. It is getting more difficult to find qualified welders.
   2. The operators must be involved!
D. Increased productivity
   1. Faster air cut speeds lead to higher duty cycles and shorter cycle times.
   2. Internal load/unload times can double a cell's output.
   3. Part manipulators can cut cycle times dramatically by reducing unproductive manual part manipulation.
   4. Changeover times can be eliminated or greatly decreased.
   5. Increased welding travel speeds reduce arc-on time.
E. Uniform, predictable quality
   1. Higher quality may eliminate prior or subsequent operations.
   2. Higher quality leads to better aesthetics, which leads to more satisfied customers, which leads to more business.

F. Operator safety

    1. Automation can provide a much safer work environment for your welders, since they can be separated from the smoke, heat, and sparks.

    2. OSHA is happier.

G. More accurate cost controls

    1. Production rates are more predictable, regardless of time of day, month, or time of year.

H. Consistent weld parameters

    1. Costly overwelding can be controlled.

    2. Consistent variables can increase weld quality.

I. Predictable material usage

    1. More accurate control of consumable parts is possible.

    2. Purchasing of welding wire and shielding gases is more predictable.

    3. Wire and gas can be purchased in bulk for greater savings.

J. Limitations of automated welding

    1. Machines do not like inconsistent parts.

    2. Machines really do not like gaps.

    3. Cleanliness of parts provides more consistent, higher quality welds.

    4. Operators must be kept informed and must have ownership of project.

    5. Ratio of welds performed by the welder to those made by the machine is important; you may not want the machine to do all of the welds.

2. Welding Automation Defined

Although robots are discussed at length in this book, there are many varieties of automated welding machinery available. Robots are by no means the only, and are often not the best, solution. A surprising percentage of automation applications are better suited for some other type of automation, usually less expensive than a robotic system.

Types of Welding Automation

Flexibility is a major advantage of robotic welding systems. What if, however, your contract is a long one, and the parts contain only simple welds? You may not want to invest in the higher cost of robots when the machines will be doing the same simple tasks for so long with no changeovers. Hard automation may be the key to decreasing your capital investment and decreasing the length of your payback period, without sacrificing much (or any) productivity. This chapter will focus on various types of welding automation available and a description of some possible applications for each type.

Machine welding

The most basic type of automation consists of some basic machine motion imparted to the welding torch or the weldment. This is commonly called "machine welding" and usually takes the form of a carriage that rides on a rack and pinion, with forward travel speeds being controlled by a knob on the carriage, as shown in Figure 2-1. The welding voltage and wire feed speed can be either controlled at

25

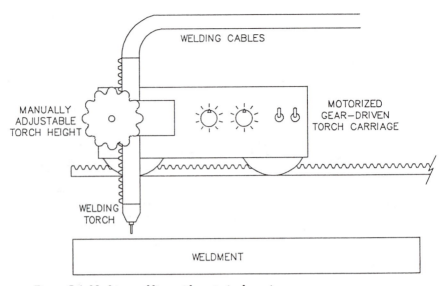

Figure 2-1. Machine welding with motorized carriage.

the welding power supply, as if the machine were being used for manual welding, or by a simple controller mounted to the carriage.

The setup of such a machine requires that a track be laid the length of the weld and that it be aligned to keep the torch in the joint as the carriage travels along. This can be time consuming, but, when compared to manual processes, the overall labor savings can be attractive. Even this simple form of automation can remove the welder from the unhealthy environment caused by the smoke and sparks and may even free him to accomplish other tasks while the machine is welding.

This particular configuration can work well for one-of-a-kind fabrications where long, continuous welds are made but not repeated. It becomes especially productive when these long welds require multiple passes. It takes some time to lay the track for the first pass, but then it is a simple adjustment to offset the torch for the subsequent passes. Common uses are in shipyards, for welding bulkheads and large plates together, and in heavy fabrication shops that make very large, unique weldments.

Machine welding can also include a setup that is common where cylindrical shapes are welded. The cylinder is placed on turning rolls, which are available for handling tremendous weights, and the welding torch and equipment are mounted to an adjustable beam. The turning rolls rotate the part at a predetermined welding speed, while the stationary torch welds the seam. This process is used for welding heavy pressure vessels, railroad tank cars, and large containers for bulk gases or liquids but is also common for welding smaller cylindrical weldments (Figure 2-2).

Hard automation

"Hard automation" refers to machines that are able to be fixtured or hard tooled for a specific application, usually involving high part volumes. If a manufacturer were to weld 500,000 flanges to pipes, all identical, then hard automation may be a good solution, such as that shown in Figure 2-3. The cost is typically a fraction of the cost of a robotic system and, because of its simple mechanical nature and precise repeatability, can easily match or exceed the quality of robotic welds.

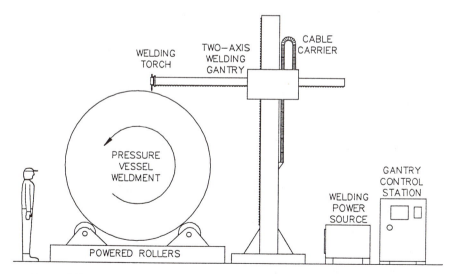

Figure 2-2. Gantry-mounted welder for welding of pressure vessels.

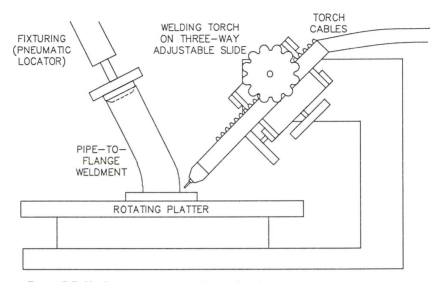

Figure 2-3. Hard automation: circumferential welder.

These machines usually consist of the following components:
- A method of moving or rotating either the torch or the part
- A fixture to locate the part(s) accurately
- A welding power supply, torch, and other peripherals
- A controller to start and stop the weld at the proper location, to control the welding speed, and perhaps for other features, such as overlapping at the end of the weld or burnback control to prevent the wire from sticking in the puddle

Such machines can usually be set up to weld various parts, but normally the parts must be very similar to each other. In the example above, the machine for welding flanges to pipes has a certain amount of adjustability built in, so that it can be easily changed to accommodate a larger diameter pipe. However, the length of pipes cannot vary too much, due to constraints in the physical size of the machine's working envelope.

A manufacturer of automotive exhaust pipes welded a short, 2.5 in. diameter pipe to a flange in very high volumes. By using hard automation with automatic ejection, they were able to complete a part about every 6 seconds. The challenge was not in meeting production requirements but in the operator's ability to keep feeding the machine parts for 8 hours per day. Other common uses include straight-line welding for joining two ends of a bracket simultaneously or a weld where the torch must follow a simple contour.

Usually, volumes must be rather high to justify such equipment. Although the cost of the equipment is low, its flexibility is limited, so the jobs on which it can be utilized may be few, especially for job-shop style work.

Flexible hard automation

To provide some additional flexibility, engineers have developed a new class of machine. Its function lies somewhere between hard and flexible automation, hence the creative title "flexible hard automation."

"Flexible" implies that the machine can be used for a variety of purposes and that changeovers are relatively quick and simple. "Hard automation" refers to the simplicity and reliability with which the machine operates after it has been set up for a particular application. Thus, we have flexible hard automation.

The difference between these two categories can be quite subtle. By increasing the working range of a hard automation machine and by adding quick release tooling for fast changeovers, and a programmable controller for changing the welding parameters by the touch of a button, you have created quite a lot of flexibility. This can allow you not only to weld parts in the same family but also any part with a similar joint type, provided it fits the general parameters of the machine. Typical applications for flexible hard automation include bicycle frames, exhaust manifolds, and metal furniture assemblies.

While hard tooled machines may have only one axis of freedom— the rotation of the part or the path of the torch—these flexible machines may have several. The welding torch may follow some contour by means of servo controlled axes or by some mechanical means. Combine this with the motion of the part, and some very sophisticated weld geometries can be welded very effectively.

Flexibility is worthless if changeovers cannot be accomplished quickly. To this end, these machines are fitted with quick change tooling, or universal tooling that may accept several part numbers with no changes at all. If the location of the weld joint cannot be consistently located, then seam tracking systems and part gaging mechanisms are available as reasonably priced options. However, as with any automated welding, it is preferable to provide component parts that need no gaging, since finding the joint before welding can increase the cycle time. By presenting accurate, consistent parts to the machine, you are building an altogether higher quality part.

A typical machine cycle works as follows. The operator first loads components into a fixture. Some manual clamping may be involved, but often the parts are automatically clamped. After loading, the operator steps back and pushes the ready button. If so equipped, automatic arc shields swing into place to shield the operator from the harmful weld glare. At the end of the typically short weld cycle, the arc shield swings open, and the part is either automatically ejected into a basket or is manually unloaded. If the machine has automatic eject, then, during the weld cycle, the operator can busy himself with getting the next set of component parts ready for loading. Such a machine can also be equipped with a rotating table on which duplicate weld fixtures are mounted. This allows the operator to load and unload at one station while the welding takes place at another station.

Flexible automation

For our purposes, "flexible automation" means robots. Throughout this book, I will use "robot" as a generic term for automation but remember that a robot may be overkill for your specific needs.

Most modern welding robots today are six-axis articulated arm manipulators. They range in size from the very small 3 kg (6.6 lb) capacity models to the large 100 kg (220 lb) and larger payload models used for spot welding and other heavy duty purposes. Almost every popular robot on the market today functions in the same basic way.

A typical robotic welding system may consist of the following components:

• Robot arm or manipulator

• Robot computer controller

• Teach pendant, a portable device attached to the controller by a cable, through which robot programs may be written, modified, and many other functions enacted, depending on the robot model

• Welding power source with wire feeder and torch

• Usually a welder interface that enables the robot controller to communicate to the welder

• Work holding fixture

• Operator's controls for starting and stopping cycles

• Safety equipment to prevent approach to the robot during program execution

Other features that may or may not be needed, depending on the specific application, include automatic torch cleaners, workpiece exchangers, workpiece manipulators, bulk welding wire dispenser, etc. A typical robot cell is shown in Figure 2-4. This cell has four welding stations, and each can have a different weld fixture mounted

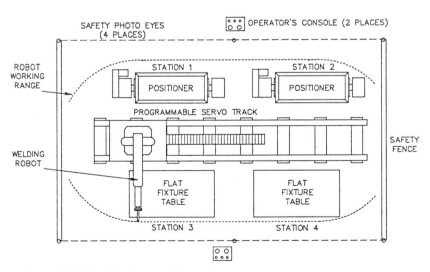

Figure 2-4. Flexible robotic welding cell.

to it. Load/unload time is internal to the robot welding cycle, and even fixture changeover can be accomplished while the robot works at other stations.

If you skim the rest of this book, then at least slow down and read this paragraph thoroughly; the success of a robotic welding installation is not guaranteed. It depends upon the following important but achievable factors.

- The proper definition of the project before the vendor quotes or makes commitments
- The technical ability of the vendor to meet the customer's real needs
- The preparedness of the customer to properly implement and prepare for automation (a good vendor can help here with advice)
- The attitude of management and of the workers who will be affected by the changes

These topics will all be discussed in detail later, but remember their importance. This is where many users have lost the battle, only because they did not know the basics.

FMS (Flexible Manufacturing System)

FMSs have allowed manufacturers to run production with the lights off. It can greatly decrease labor, but the cost can be overwhelming. You must have the correct application to justify the purchase of an FMS.

True applications for FMS (in our case FWS, or flexible *welding* system) are few and far between and very highly specialized in nature. When applied to welding, flexible manufacturing can take on some truly innovative and unique forms, but the purpose is always the same; to weld lot sizes of one as easily as 100. To utilize this technology properly, families of parts must be identified that can fit within the operating parameters of available machinery.

One type of configuration for FWS is shown in Figure 2-5 and Figure 2-6. The fixture load/unload positions give the operator(s) a choice as to which part will be loaded next. If one of each subassembly is required, then each of the several fixtures can be cycled through the robotic welding station in the proper order, but if

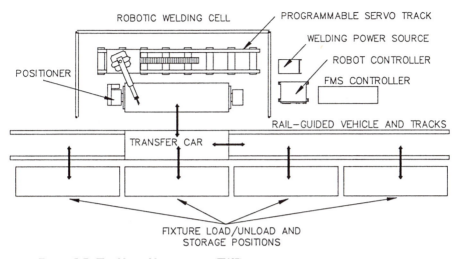

Figure 2-5. Flexible welding system (FWS).

several of one assembly is required, then that can be accomplished just as easily. The rail-guided vehicle is instructed to retrieve a particular pallet or can be programmed to retrieve each one in order automatically, for lights-out manufacturing. The positioner in the robot cell must be equipped with automatic coupling devices in order to receive the pallet and lock it into place for manipulation, and the

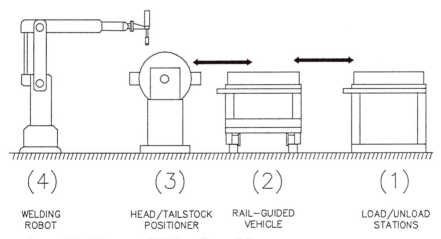

Figure 2-6. End view of FWS from Figure 2-5.

rail-guided vehicle must be equipped with a storage and retrieval mechanism on top that can effectively transfer pallets from the load/ unload positions to the welding positioner.

The FWS can be designed once the family of parts is identified. If the parts to be welded require no positioning during the weld cycle, then a system that presents parts to the robot on a flat pallet will be sufficient. If parts require positioning, then an FWS can be designed that automatically shuttles parts onto a workpiece positioner for welding.

The advantages to FWS technology are numerous. The most obvious, already mentioned, is the ability to weld small lot sizes profitably. One way this is accomplished is by using coded pallets. A pallet can be coded by any number of means, according to the weld fixture that is mounted on it. When the pallet is shuttled into the welding station, the code is read by some type of sensor, the correct program is downloaded into the robot memory, and the robot welds the part—all automatically. It does not matter how many of a particular part number are shuttled through, or in what order they are presented to the robot.

A typical coding method involves the use of proximity switches. A code is attached to the fixture base plate, consisting of a plate with a certain hole pattern drilled in it or perhaps a row of bolts or machined bosses providing a unique pattern of raised surfaces. On the surface to which the fixture will be mounted—a transportable pallet or the faceplate of a positioner—is mounted a row of proximity switches. When the fixture is in place, the robot controller is instructed to read the code. Depending on the binary input received from the row of proximity switches, a unique welding program will be automatically downloaded into the robot's computer from a mass storage device or hard disc drive. Thus, each fixture can be unique, and welding lot sizes of one become a reality.

FWS and Just-In-Time (JIT) Manufacturing

Beyond the buzzword that it has become, JIT can save you a lot of money, and robots are very capable of providing it, in the right circumstances. FWS is a technology specifically designed for JIT manufacturing. A robotic welding cell has the capability to weld one of each subassembly, providing your assembly department with just enough components to build a complete widget. If the subassemblies then need to be welded into a final assembly, the robot may even do that. Raw materials can be fed in one end of the FWS, and a finished component comes out the other. This is true JIT manufacturing and allows you to respond to your customers like never before.

If you sell your products through distributors, think of the advantages, if you are not already doing so, of building per their orders. Whether hot-shot "Distributor A" has placed orders for 300 widgets, or small-time "Distributor B" needs only 3, you can fill both orders just as easily and profitably.

Another advantage, although not so commonly implemented, is the ability to do lights-out manufacturing. Many FWS have "pallet pools" or queuing areas that allow operators to get ahead of the robot by loading up fixtures faster than the robot can weld them. Obviously, this is only possible with assemblies requiring longer welding cycle times. By loading up at least one shift worth of robotic welding work in the pallet pool, the robot can work alone with no operator intervention until all the welding backlog is completed.

For true JIT manufacturing, changeover time must be eliminated in terms of interrupting the actual welding operation. The pallet pool not only provides a queue for the robot but also provides the ability to change fixtures on the fly, with no robot downtime. The best way to accomplish this is to attach the correct fixture to the pallet off-line, apply the correct code to the pallet, then simply set the pallet into the system at any point. When the pallet reaches the robot, it is correctly identified and welded.

FWS is not for everybody. If, however, you weld lots of part numbers in smaller lot sizes, and the parts always repeat, say, every week or every month, and if you want virtually to eliminate robot downtime due to setup and changeover, then FWS may be worth pursuing.

A few final words of advice about FWS.

● There are not many suppliers out there that can do it effectively, so choose wisely.

- It can be very expensive but can provide unparalleled flexibility.
- Be sure to investigate mass memory; many robot systems do not have enough RAM to handle all of that information effectively.
- FWS can be used for long high-volume runs, but that requires a lot of duplication of fixtures, which can become very costly.

More on Flexible Automation

Robotic welding is the hottest topic in welding circles for a reason. It is fun. Not everyone has one, so there is some novelty involved in owning one. You realize, however, that these are not the reasons to invest in robotics. The reason for all of the hoopla is flexibility.

Flexibility

A robot can do practically anything a dedicated machine can do and more. It can be reprogrammed very easily for other jobs and can run completely different products simultaneously. Robots can weld large and small parts, can work with manipulators to position the weldments, and can be put on tracks, mounted overhead, or on a gantry. Robots give companies opportunities to quote projects they may not otherwise have been able to. Job shops can set up several workstations around a robot, leaving certain stations permanently set up on those few long-term jobs, while changing other workstations over regularly. If you have the guts, you might just try buying a robot, with no job in house to justify it. I have seen it done, and it works. Most customers, including yours, would prefer to do business with a vendor who will robotically weld their parts.

Once written, programs can be stored on various media for later use, and the welds will look exactly like they did the last time that part number was run, provided the system is designed and built properly. These storage media include the following:

- Cassette tapes, which are becoming more outmoded as faster and more flexible disc drives take over. Most cassette tape systems cannot deliver just one program to memory, but the robot must read all the information that is on the tape. This can be frustrating, especially when you need to make a change to only one of the programs on the tape. The alternative to storing many programs on one tape is to use one tape per program, but, of course, this necessitates having a potentially huge number of cassette tapes available.

- Floppy disc drives are the most popular source of memory storage. This is due to their flexibility and large memory capacities. Standard high-density computer discs can be used (some systems require only low-density discs), which provide a lot of storage space, and single programs can be uploaded and downloaded for convenience.

- Hard disc drives provide much larger memory reserves and can access information much more quickly than a floppy disc drive. These are more expensive but are well worth the cost, since all of your active programs—depending on their number and lengths—can be stored on the hard drive. You would then use floppy discs only for backup purposes. This saves your floppies from the harsh atmosphere of welding and keeps them in good condition, since they are not accessed several times per day. Hard disc drives with very short access times are available and can be used "on-line." This means that, if you have a very long welding program that is too large to fit into robot memory, then the robot first downloads as much of the program as it can into memory. When that portion of the program has been executed, then the hard drive immediately loads more of the program into the robot's memory, until the entire program has been executed. All of these factors contribute to the amazing flexibility of welding robots.

Looking to the future

Robotics provide for your future like no other type of automation. Programmability ensures the ability to do any job that may fit within the practical limitations of the robot cell. Multiple axis movements allow you to reach a large percentage of the welds on a particular assembly, and the future additions of servo tracks, positioners, and other peripherals guarantee the ability to expand into other types of jobs and products.

Most robotic cells today are easily retrofitted in the field with additional equipment. Just remember that it will cost more than if the robot had been originally purchased with the equipment. If you are a job shop, then be sure to consider what a positioner can do for you. If you happen to be quoting a job that requires positioning during the weld cycle, you do not want to regret not having purchased a positioner.

Are positioners necessary? One company I visited had purchased several welding robots but decided to save some money in the process. Although the weldments required positioning, they decided to purchase the robots without positioners and add manually operated trunnions to the cell. When the robot finished a particular segment of the program, it would stop and wait for the operator to turn the fixture over manually, lock it into place, and push the go button again. It saved them some money, to be sure, but the operator was also involved with other tasks in the cell and was not always there when the robot stopped. Since the robot cycle time was not very long, the 30 seconds that the robot had to sometimes wait was a significant portion of the overall cycle time. Could they have justified the additional cost of positioners? Probably. Do not buy the best price, buy the right solution. The two can be mutually exclusive.

Various types of workpiece manipulators, or positioners are designed to work in conjunction with welding robots to increase the effectiveness of the robot's working range. By bringing the welds to the robot with the aid of a manipulator, a higher percentage of welds on a particular weldment can be reached, thereby increasing the efficiency of the robotic welding cell.

Welding positioner design and function

Following are descriptions and drawings of various types of positioners used, usually in conjunction with welding robots, to manipulate parts during welding. These represent only the basic types of positioners available. The configurations and functions of welding positioners are limited only by the builder's abilities and imagination. The key is to bring the welds to the robot, making it possible for the robot to reach a higher percentage of the welds on a particular weldment. This can dramatically increase the efficiency of the robotic working cell.

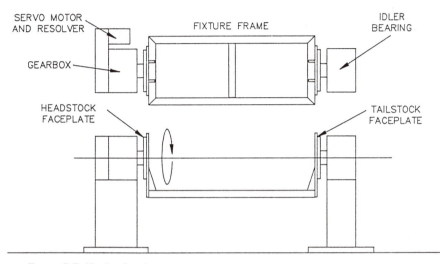

Figure 2-7. Head/tailstock positioner.

The head/tailstock positioner, also known as a trunnion, is quite flexible, even though it possesses only one axis of rotation. The headstock is powered by either an indexer, which can stop in 4, 8, or 12 positions (or almost any number of positions required), or by a programmable servo motor and resolver. The tailstock is supported by an idler bearing.

Typically, a frame is supplied between the headstock and tailstock on which fixtures are mounted, but often the weldment itself can act as the framework, attaching the headstock to the tailstock. By excluding the frame, the robot is freer to access all sides of the weldment.

A disadvantage of such positioners is that it can be difficult for a fixed robot to reach inside certain objects such as backhoe buckets or tanks, especially when turned away from the robot. In such cases, the robot's access to the part can be improved by placing the robot on a track or shuttle. Head/tailstock positioners are most commonly available in capacities up to around 5000 pounds, but, of course, there really is no practical limitation to the size of a positioner other than its cost.

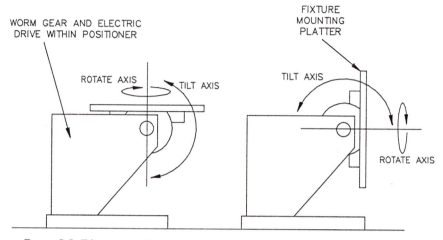

Figure 2-8. Tilt/rotate positioner.

The tilt/rotate positioner features two axes of rotation, again either indexing or servo-controlled, that together provide a high degree of flexibility. The platter rotates while the second axis tilts the part, usually from a flat position through a range of 90 degrees or more. Drive motors and gearboxes are located within the body of the positioner itself.

Welding fixtures are mounted to the rotating platter, so the fixtures themselves will differ from those designed for use on head/tailstock positioners. Although more flexible in its movements, the tilt/rotate positioner does not allow access through the bottom of the fixture as with the head/tailstock design.

The distance of the weldment center of gravity (CG) from the platter is an important factor, since there is no tailstock to help support the weight of the part and fixture. The entire load is cantilevered from the rotating platter. This requires a very strong drive unit and gearbox, which, on the larger capacity positioners, can add considerably to the cost. Some weldments, such as very long and heavy structures, simply should not be manipulated by the tilt/rotate positioner, since the cantilevered load causes the CG of the part and fixture to be extended too far from the platter. This particular positioner is very sensitive to CG location.

A disadvantage with the tilt/rotate design is that, during tilting, the weldment itself moves a significant distance through space as its CG is displaced. This may require that the robot be mounted on a track or shuttle in order to reach the weldment in both the upright position as well as the tilted position.

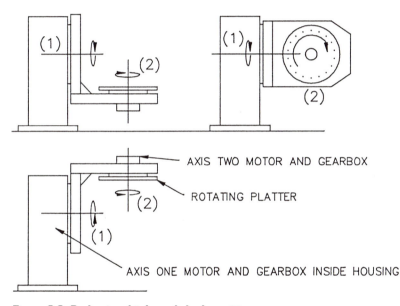

Figure 2-9. Dual-axis orbital, or skyhook, positioner.

The "orbital," or "skyhook," positioner also utilizes two axes of motion but is more flexible than the tilt/rotate positioner. The arm rotates around the horizontal axis (1), and the platter rotates around axis (2). The additional flexibility is due to the relative motion of the two axes; the CG of the weldment does not move significantly but stays in the same area in space during manipulation. The part does not move outside the working range of the robot during positioning, so the robot is used more efficiently. This can eliminate the need for a robot track, thus eliminating significant expense.

Cantilevered loads again require sturdy drives, but, since the load CG can be more easily maintained at the centerline of rotation, the stress on drives and bearings can be reduced. Again, this positioner is not recommended for long parts or for weldments that require the CG to be a significant distance from the platter.

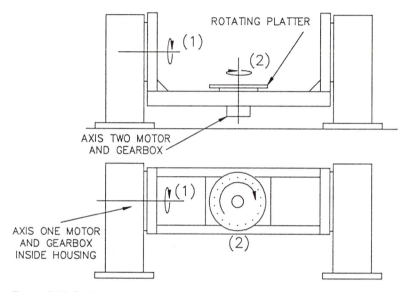

Figure 2-10. Dual-axis drop-center positioner.

The "drop-center" positioner is very similar to the orbital positioner in its movements, since two axes of motion are utilized in much the same way. However, a tailstock provides added stability and greater load-bearing capacities. The motions are identical, with one horizontal axis providing tilting motion and a rotating platter providing rotation of the load. As with the orbital, the load CG does not move through space significantly, so the robot's working range is used efficiently.

Flexibility is slightly limited as compared to the orbital positioner, since the tailstock can limit the robot's access to the weldment. This positioner is typically used in applications where large, long, or heavy weldments would place too large a cantilevered load on an orbital or tilt/rotate positioner. This positioner is really a combination of the head/tailstock and the orbital positioners and is normally used in applications where the combined weight of the fixture and weldment require the extra support of the tailstock.

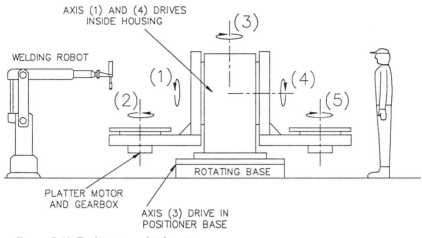

Figure 2-11. Dual-station orbital positioner.

By placing two positioners on a rotating base, the load/unload time can be made internal to the robot's welding time. This drawing shows such a positioner, with two orbitals mounted to a common base that rotates 180 degrees. The operator works safely outside the robot's working range, loading and reloading the fixtures. When ready, the operator instructs the system that the parts have been loaded. The robot then instructs the positioner to rotate, presenting the loaded, unwelded fixture to the robot, and the welded parts to the operator for unloading.

Although more costly, this is one of the few ways to safeguard such a system. The alternative is to place two individual orbitals on the floor within the working range of the robot. This causes the operator always to be within the robot's working range, in danger of being struck by the robot while reloading fixtures. In addition to increased safety, the configuration shown may prevent the need for an expensive robot servo track and will decrease the amount of walking the operator must do.

This positioner is sometimes called a "five-axis" positioner, since it has two axes per side, plus an additional rotating base axis.

Although accurate, this terminology can be misleading, since the rotating base often is no help in actually welding the parts. However, a rotating base can be designed to stop in 90 degree intervals, and these extra positions can be used for additional flexibility in reaching welds with the robot.

Most other types of positioners can be mounted on a rotating base in a similar manner. Common designs include dual head/tailstocks and dual orbitals, but some welding automation suppliers can supply almost any combination you need to optimize your investment.

Fixturing Issues

Fixturing is often the most neglected part of an automated welding system. Many users of automation prefer to build their own fixtures to save money. While it is true that the tooling is often a significant part of the total cost, it may be better to buy your fixtures from the vendor along with the machine. You may be expert fixture builders, but you may not know how fixtures should work in conjunction with machines. A machine builder will know exactly where the welding torch can and cannot reach, how much clearance is needed for a robotic torch bracket, and how to assure that the part can be unloaded after it is welded. Many more constraints exist for automated welding fixtures than for those designed for manual welding. At the very least, purchase the first set of fixtures from the supplier. Then you can learn how it is done and build additional fixtures yourself when they are required.

Fixtures for automated welding see more abuse than those used for manual welding. If a robot has a cycle time of 1 minute, then it is possible for that fixture to see hundreds of parts in one shift. Since a machine's duty cycle can be three times that of a welder, the wear and tear on the fixture will be proportionally higher. This will wear down even hardened steel locators in time. Locators, clamps, and anything on a fixture that moves should be replaceable, since the fixture will see such high duty cycles. New issues arise when designing fixtures for automation, so you need to work with someone with experience.

One factor that is often overlooked is the requirement of the welding fixture to withstand dynamic loading. If the part to be welded weighs 1000 pounds, and the fixture weighs 1000 pounds, many companies will design the positioner to withstand loads of 2000 pounds, plus a safety factor. However, often they do not take into account the dynamic loading of the positioner, which occurs during loading or unloading of the parts. If an operator drops hundreds of pounds of parts into a fixture, the positioner will see loads in excess of its 2000-pound design limit. If the operator tries to yank a "stuck" part out of the fixture with a 2-ton crane, the positioner can receive stresses for which it was not designed.

The integrity of the fixtures and positioner is especially important when considering dual-ended positioners. If the end at which the operator is loading and unloading is seeing these stresses from loading and unloading, and the other side—the side which is at the robot being welded on—is vibrating and shaking from the pounding, then weld quality will suffer. Be sure that the equipment you purchase is well built, and such load-induced dynamic stresses—which are often unavoidable—will not interfere with the normal operation of the system.

Seam Finding and Tracking

The ability to find and track seams is one of the fastest growing segments of the automated welding market. It seems that everyone is preoccupied with finding a better, faster, and cheaper way to get around the core issue—inconsistent parts. Now, in some cases, it just is not possible to prepare or fixture parts properly so that the weld seam is always in exactly the right spot. In these cases, some type of adaptation is needed. I simply want to make it clear that your first option should be to correct the problem, not to throw money at it. To search for a weld joint adds time to your weld cycle, decreasing the productivity of the cell. These options cost money, which can lengthen your payback or cause you to cut corners elsewhere. The following discussion assumes that the problem just cannot be solved.

In this case, there are some excellent methods available for locating and following weld joints.

Mechanical gaging

Hard automation can be fit with mechanical gaging devices that automatically find the weld joint. A locator is indexed against the weld joint, while the welding torch is locked into the correct relative position. The locator is retracted, and the welding commences. This is a common application for circumferential welding, for instance, when welding around a tube.

Real-time mechanical followers are also common. The simplest type is made up of simple rollers that ride along the material being welded. The rollers cause the carriage to rise and fall mechanically, keeping the torch properly located in the joint. The advantages to such a system are its fail-proof method of operation and low cost. The main disadvantage is the limited joint types with which it functions well.

Many utilize a different, but very effective, theory: a follower rides in the joint ahead of the arc, sensing changes in direction during welding. Information of various sorts is fed back to a controller that automatically corrects the welding path. The torch motion could be controlled, for instance, by air or by servo motors that react to this feedback. These are a little higher priced, but still very reasonable when compared to vision. Again, the use of such followers is often limited to fillet joints and lap joints in heavy material. There must be a groove to follow in order for this technology to function properly.

Tactile sensing

Tactile sensing, or touch sensing, is the ability of a robot to touch the weldment with the wire or some other device and find the joint. This gives the robot the ability to tolerate a weld joint whose location is floating around from piece to piece.

This is how a basic tactile sensing system is used to locate a wandering weld joint. A voltage, usually around 40 volts, but often as high as 200 or 400 volts, is superimposed on the welding wire, nozzle, or some probe at the end of the robot arm. The robot is programmed, at every cycle, to move slowly toward the part until it touches. Searches only need to be done on those welds that are not capable of being accurately located—not on every weld.

Please refer to Figure 2-12. Position 1 is a home position or the previous position in a program. Position 2 is the approach point, which is programmed parallel to the bottom horizontal plate. As the robot moves toward the part, it is waiting for a signal that it has been grounded against the part, as shown at position 3. The grounding of the voltage to the workpiece signals the robot to store those X, Y, and Z coordinates in memory. The robot immediately approaches position 4, using the newly calculated offset to position the welding torch precisely in the joint.

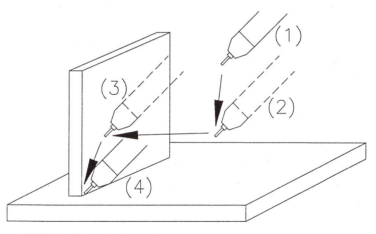

Figure 2-12. One-dimensional touch sense routine.

Multiple dimensions can be searched to locate the weld joint exactly. This is done by simply completing a search in each dimension and storing all of these coordinates in memory. Figure 2-13 shows how a two-dimensional search is done. Position 1 is a home point. Position 2 is an approach point for the first search. Position 3 indicates where the wire touches the part and stores the first coordinate. At this point, each subsequent search will also

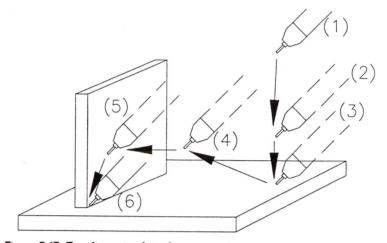

Figure 2-13. Two-dimensional touch sense routine.

contain the offsets discovered by this first touch. The second search begins at position 4. Position 5 is the second touch, and the robot controller stores this second coordinate. At this time, both an X coordinate component and a Z coordinate component are combined to determine where the weld joint is, which is position 6. A third dimension may also be searched if necessary.

If the part cannot be touched for some reason, other types of sensors may be used. The robot is normally just looking for a digital signal, so proximity switches and other such sensors can work well with the same tactile sensing circuitry. The only limitation to these options is that some sensors cannot stand up in the welding environment, especially with the high duty cycles that robots generate. Be wary when scrutinizing sensor technologies that have not been fully proven in a welding atmosphere.

Touch sensing is a proven technology that has been around for years and can work very well in the right circumstances. It adds time to the robot cycle, but often there may be no other choice. However, there is a major limitation of tactile sensing: it is not a real-time process. The robot searches the end of the weld and stores that point in memory. Then the beginning of the weld is searched, and that point is also stored. If a third point is needed, then a third search is made.

The computer then instructs the robot to weld from the start point to the previously stored end point, but the weld must lie along a predictable path, since no tracking is taking place during the weld. Thus, thermal distortion can be a problem on longer welds and thinner metals. Since the tactile sensing establishes the locations of the weld start point and the end point prior to welding, any movement of the part during welding cannot be compensated for. Thermal distortion can actually move the parts during welding, so any stored end points will become meaningless as the weld progresses. However, this is not true in most cases, since proper fixturing should keep parts from distorting during welding.

Another disadvantage with tactile sensing is the time it takes to complete a search. Search times vary from about 2 to 5 seconds per search. This means that if you require a three-dimensional verification, it could add up to 15 seconds per search to your cycle time. A technology is readily available to cut down on this additional time by limiting your searches, in many cases, to one or maybe two dimensions. This technology is known as through-arc seam tracking or, more commonly, through-arc tracking.

Through-arc tracking

Through-arc tracking is also a common, proven technology that is normally offered as an optional accessory, since many jobs do not require any sensing or tracking at all. This is a real time process, meaning the tracking takes place during welding, so any unexpected curvature of the joint or movement of the part due to thermal distortion while welding is compensated for by the tracking system. Through-arc tracking is usually used in conjunction with tactile sensing. The beginning of the weld joint is located through tactile sensing, then the through-arc tracking takes over from there. This can cut down your cycle time considerably when compared to doing three-dimensional searches.

To decrease the cycle time even further, through-arc tracking can, in some cases, be used without tactile sensing. If your weld fillet size is 3/8 in., and you know that the weld joint will never vary in location more than ⅛ in. in either direction, then you may not need to locate the weld start with tactile sensing. The robot will simply start welding, and, if the arc is close enough to the side walls of the fillet for the through-arc sensor to get a good reading, then the robot will "find" the joint quickly, even if the weld joint was out of location by ⅛ in. The arc will be directed into the center of the joint via the through-arc sensor, and you may not be able visually to tell that any correction was made.

Figure 2-14 shows how through-arc tracking works. The basis of operation is the fact that stickout—the distance between the end of the contact tip and the workpiece—is directly proportional to amperage in a constant-voltage welding system, and arc length is directly proportional to arc voltage. The longer the arc, the higher the voltage.

First of all, it is necessary that weaving is used, which is a function built into most welding robot controllers. The robot software continuously monitors the arc voltage and current during weaving. As the arc approaches each side of the joint, the arc length and tip-to-work distance shortens, which changes the welding voltage and current. The computer senses these changes and strives to keep these values equal on the left and right sides of the joint by automatically

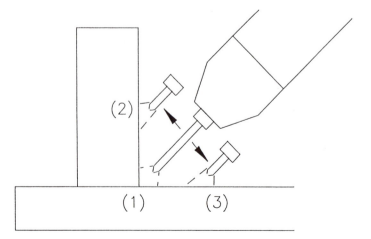

Figure 2-14. Through-arc seam tracking technique.

offsetting the preprogrammed path of the robot. This keeps the arc traveling down the center of the weld joint.

The torch-to-work distance is maintained by measuring the voltage and current as the arc moves past the center of the weld where the standoff is longest. The computer then strives to maintain a preset value at the center in order to keep the standoff distance constant. This function can be used alone (without the left/right correction) for keeping the standoff distance constant while not weaving—for example, on a butt joint or lap joint where weaving might not be necessary.

One key disadvantage with through-arc tracking is that weaving must be used, which makes it impractical for thin-gage metals. Weaving is usually not possible on thin-gage metals, since fillet sizes are normally too small to allow weaving. If larger fillets are used, then too much heat is generated, and the weld burns through the plate. The smallest practical fillet joint with which through-arc tracking can be used is about $1/4$ in., or perhaps $3/16$ in. depending on the circumstances.

Laser tracking

Lasers have become quite useful for finding and tracking weld joints. Although their use is not nearly as widespread as tactile sensing and through-arc seam tracking, this technology has proven effective when installed in the correct application. Therein lies the key to successful use of lasers—the correct application.

Laser trackers are typically used for welding lap welds in thin-gage materials. The laser can "see" the lap and, through sophisticated algorithms, can compute corrections in the robot's path in real time. This also holds true for butt welds, unless the fitup is so tight that the gap cannot be seen by the laser. Of course, fillet joints can be easily tracked by the laser, although if fillets are the only type of joint to be welded, less expensive solutions might be used effectively.

The laser can be used as a seam finder, so tactile sensing is not necessary to find the beginning of the joint. By scanning in a preprogrammed pattern, the system can locate the beginning of the joint, then can begin welding while tracking in real time. This provides for some good flexibility, since all functions are contained in the laser software itself.

During operation, the tracking system scans back and forth across the weld joint with a laser, while gathering imaging information (Figure 2-15). The corrections are sent to the robot controller, and the robot is instructed to follow the joint. If clamps are encountered that fall within the sight of the laser, or if the weld joint is close to another edge that might confuse the laser as to which edge to follow, certain parameters can be programmed into the system to cause the computer to ignore these extraneous inputs. This allows further flexibility as to where and what type of joints can be effectively tracked. Laser systems can also provide a certain amount of "adaptive fill." Based on gaps that the laser system detects, the robot can be instructed automatically to call up certain preprogrammed weld parameters to address the changing gap conditions. Either more or less wire feed and more or less voltage will be supplied to fill the gap appropriately.

Limitations with most lasers include the limited weld joint accessibility, due to the amount of hardware around the welding

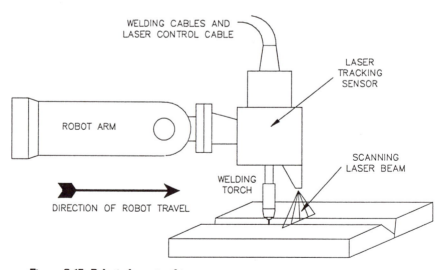

Figure 2-15. Robotic laser tracking system.

torch. Another drawback is the cost. Laser seam finders are one of the more expensive means of tracking joints. One more limitation is derived from the function of the tracking system itself. While tracking, the laser must scan about 1 in. ahead of the weld, due to all the visual interference from the arc, spatter, and smoke. Although rarely a problem, this can cause weld defects, due to the purposeful 1-in. time lag in sending corrections to the robot arm.

There are laser seam finding systems available that do not have the ability to track joints. These systems work much like tactile sensing; they merely locate the start and end points of a weld. The most useful application for these is in finding lap joints in sheet metal. This is very difficult with tactile sensing, since there is nothing to touch. A lap weld on 0.030 in. material would not have an edge large enough for tactile sensing to find, but a laser can "see" this joint easily. Although more costly than tactile sensing, laser seam finders are more reasonable than real-time laser trackers.

Robot vision

Robot vision has been a buzzword, since it conjures up images of the high-tech nature of robots and the mythical feeling of invulnerability that comes from owning a welding robot that can see. The truth is that true vision applications are very few and far between. My opinion is this—if your weldments are in such poor shape and so devoid of repeatability that vision is needed in order to weld them together, then you have much larger problems to address. The cost of vision is very high, and systems are comparatively difficult to run and maintain, requiring the user to keep a programming expert available at all times. The automated welding system, which should be viewed as an extension of your welders' skills, becomes a technological marvel that can probably no longer be operated by those with the welding knowledge.

Before you completely write off vision as a solution to your problem, let me discuss a few features that vision provides that are not possible, or at least not so effective and flexible, with other tracking systems. These features include adaptive fill, typically much more flexible and versatile than the adaptive functions available with laser systems. This gives the robot the ability to see the weld joint and, based on calculations of the volume of weld metal needed to fill the joint, it can change weld parameters (weld voltage, wire feed speed, travel speed, etc.) to provide just the right amount of filler. Artificially intelligent decisions can also be made by such a system. All of this adds up to quite high costs and complexity, so, before considering vision, be sure you have thoroughly checked out your alternatives.

Avoiding seam finding and tracking

This discussion on seam finding and tracking would not be complete without the following point being strongly stressed: your first goal should be to avoid tracking and seam finding. Make whatever changes are necessary in your component parts and fixturing, so that no form of adaptation is needed to locate the weld joints. These options that we have been discussing only add cost and complexity to the welding system and can add precious minutes to your cycle times. You have purchased automation in order to provide higher quality and higher productivity. Do not lug it down with tracking systems, and do not extend your payback period with fancy options unless it is necessary. Remember, sometimes it is necessary, but you should never immediately default to using such options unless you well know the reasons why.

Summary

In chapter 2, we have discussed the various types of automation, as well as some of the common options available and how they function. If you understand this chapter thoroughly, then you have a good grasp on what automation can do for you and what type automation may best fit your needs.

A. Introduction

    1. Robots are not necessarily the best solution.

B. Types of Welding Automation

   1. Machine welding

      a. Usually a very simple motion imparted to the welding torch, such as a linear track with rack-and-pinion drive or a boom-mounted torch with rollers to rotate weldment during welding.

      b. Common applications include:

        (1) Long, straight welds on plate.

        (2) Oil tanks, ship hulls, girders, fabricated I-beams.

        (3) Circumferential and longitudinal welds of large and small cylinders, pressure vessels, tanks, containers, etc.

   2. Hard automation.

      a. Machines fixtured for a particular application. Usually includes a controller, welding apparatus, and hard tooling, sometimes easily interchangeable to provide some flexibility.

      b. Common applications include:

        (1) Tube-to-flange welding.

        (2) Simple brackets and other attachments.

        (3) Tube-to-tube joints.

        (4) Plasma cutting simple contours.

   3. Flexible hard automation.

      a. Hard automation that has been upgraded with additional torch motions, easily interchangeable fixtures, more flexible controls, or more flexible working range.

      b. Common applications include:

        (1) Complex tube-to-tube joints.

        (2) Multiple torch welds with complex contours.

        (3) Exhaust manifolds, furniture parts, bicycle frames.

      c. Changeovers should be quick and easy; controls should allow fast editing or resetting of weld parameters.

   4. Flexible automation (robotics)

      a. Successful installations depend on:

        (1) Proper project definition.

(2) Vendor's technical abilities.

(3) End user's preparedness.

(4) Attitudes of management and shop personnel.

5. Flexible manufacturing systems (FMS)

   a. Lights-out welding is possible.

   b. FMS applications are highly specialized.

   c. Small lot sizes are welded as efficiently as large batches.

6. Just-in-time manufacturing (JIT)

   a. Flexible Welding System (FWS) can eliminate inventory and decrease lead times.

   b. Changeover time and load/unload can be eliminated completely.

   c. FWS is complex and difficult to do correctly. Find the correct supplier.

C. More on flexible automation (robotics)

1. Robots provide unmatched flexibility.

2. Robots can help job shops bid jobs they might not otherwise be able to bid.

3. Robots provide very high duty cycles.

4. Robots provide future flexibility and freedom.

5. Robots can work with many types of peripheral equipment, such as positioners and tracks.

D. Fixturing issues

1. Fixturing is often neglected.

2. The automation supplier is normally the most qualified party to provide welding fixtures.

3. Automated welding fixtures see higher duty cycles, therefore are more abused than manual fixtures. They should be built accordingly.

E. Seam finding and tracking.

1. The first line of attack should be to eliminate the need for tracking and sensing.

2. Mechanical gauging.

   a. Mechanical followers indicate the location of the weld joint and provide feedback for torch path corrections.

3. Tactile sensing
   a. The robot touches the part to find its location, then offsets the path accordingly.
   b. Tactile sensing is not a real time process.
   c. Search routines add to the cycle time.
4. Through-arc tracking
   a. By monitoring voltage and current at the welding arc, the robot strives to keep the torch in the joint.
   b. This is a real time process.
   c. Weaving is required for proper function.
   d. Through-arc tracking is often used in conjunction with touch sensing.
5. Laser tracking
   a. By "viewing" the weld joint with a scanning laser beam, the controller is able to identify the location of the joint and offset the weld path accordingly.
   b. Laser tracking is a real time process but can also be used for finding seams.
   c. Some systems provide adaptive fill functions.
6. Vision
   a. Vision can be extremely flexible and profitable for just the right application.
   b. Vision can be extremely expensive and difficult to learn, operate, and maintain.
   c. Be careful when considering vision. Truly effective applications are quite uncommon.

3. Implementation Strategies

You don't wake up in the morning and tell yourself, "Today I will buy a welding robot." At least you should not. Automated welding is a strategy for better serving your customers, winning orders away from your competitors, increasing quality, and cutting costs. It is a war out there, and, in any war, strategy is important. There are a few right ways and many wrong ways to get into the automated welding business, and this chapter will focus on the right things to do to optimize your investment, to maximize your savings, and to minimize the risks inherent in any capital equipment purchase.

Chapter 2 described the various types of automation available and briefly discussed some applications in which they are commonly utilized. We will now discuss applications in more detail, since this is the foundation upon which a successful welding cell will be designed.

Define Your Application

In order to understand fully how you can utilize welding automation, we must first carefully examine your specific welding application. This will allow us to:

• Determine the appropriate welding process
• Determine the best type of automation available for your job
• Determine how such automation may affect other plant production processes

Weight and size of weldment

The first of many criteria you will use to define your particular application is the physical size of the item that you want to weld. Microwelding applications may include:

- Electrical leads and wiring terminals
- Certain medical instruments and biotechnical mechanisms
- Electronic control components
- Critical edge build-up, such as in rebuilding of turbine engine blades

Such tiny arc welds can be effectively automated with highly accurate, sophisticated welding apparatuses that are commonly available. Critical control is provided first by advanced welding power supplies, capable of controlling welding voltage and amperage to diminutive degrees, even to the hundredths of a volt. Automatic voltage controls, which automatically adjust the arc length in real time, keep the arc parameters within strict tolerances, thus ensuring almost unbelievable weld quality and consistency.

Micro spot welding machinery is also available that is capable of welding extremely thin or fragile sheets and wires, and such welding can be automated by the addition of small, highly accurate transfer mechanisms that convey the weldments through the welding station. Automatic feeders, such as vibratory feeders, might complete this highly profitable, hands-off welding package. Although resistance welding is a very commonly automated welding process, for the purposes of this book, only arc welding systems will be discussed.

The next step up in size might accurately be described as "small" weldments, which may include the following items:
- Household appliance components
- Power tool components
- A wide variety of automotive components, such as airbag parts, brake pedals, window handles, various and sundry brackets, seat latches, safety belt parts, and steering wheels
- Boat jacks
- Handguns and rifle parts
- Hand and garden tools
- Gages and instruments

These small parts can be very effectively automated either by robotics or by hard automation. These can be especially profitable, since setup and material handling time can be excessive as compared to actual weld time. If the handling can be done internally to the welding through multiple fixtures and multiple weld stations, then returns on such investments can be very attractive.

Medium-sized weldments are much more common and make up a large percentage of applications for automated welding. These may include:

- Motorcycle parts
- Automotive suspensions
- Exercise equipment
- Food service equipment, such as deep fat fryers and stainless steel sinks
- Metal furniture
- Frames for personal computers
- Home appliances
- Lawnmower decks
- Aluminum boat canopies
- Snowmobile bulkheads and skis
- Metal shelves and racks
- Exhaust pipes, mufflers, and catalytic converters
- Automotive seats
- Automotive wheels
- Air compressors and other pressurized tanks
- Bicycle frames

All of these can be ideal applications for automated welding, and, in fact, all of these are applications in which automated welding is being utilized today. Remember that several factors determine which type of automation is best suited for a specific application. Later in this chapter, these other factors will be discussed in detail. The important characteristic about this class of weldments is that most of them can be easily handled by people, without the need for a hoist or crane. This factor has some effect on the design and structure of a welding system.

Where robotic welding is concerned, these "medium-sized" weldments fall into the most competitive category of machinery. There are a plethora of small robots available in the marketplace, most of which will do well with such jobs. Factors to consider in choosing among these numerous suppliers include the welding technology available in the welding power sources, the programmability of the robot, the flexibility of the robot arm, the processing speed and software features, and—perhaps most importantly—the organization behind the equipment. Any machine will eventually break down. How these service situations are handled is of utmost importance.

Larger weldments present slightly more of a challenge. There are only certain equipment builders that can provide effective automated welding systems for some of these large, heavy weldments. Many who are not capable may still try to design a solution, with disastrous results. Software capabilities, hardware integrity and design, and commitment to support after the sale are very important considerations when contemplating more difficult applications. These heavy-duty jobs might include:

- Agricultural equipment, such as tractors, harvesting and planting equipment, fertilizer and pesticide applicators, and similar equipment
- Utility trucks and booms
- Ladders and scaffolding
- Car bodies
- Truck trailers
- Truck bolsters
- Ski lift towers and lift chairs
- Boat trailers
- Road construction equipment
- Logging machinery
- Rail cars and tank cars
- Rocket engines
- Steel building trusses
- Fork trucks and fork truck attachments
- Gravel conveyors
- Rock crushers

Such weldments usually require positioning during welding, but only a few companies provide positioners with such high weight capacities. Also, only a few companies are able to move the robot around in such a way as to reach a significant percentage of the welds. With such large parts, simply positioning the weldment may not be enough; the robot must be taken to where the welding is. This requires sophisticated, and usually costly, gantries, overhead mounts, and servo tracks.

Larger weldments usually have long, complex welding programs. Such programs put great demands on robot software and interactive programming hardware. Unless the robot controller and other system controls are designed for longer, more complex programs, the system could well turn into a costly mistake. Never fear, however, just do your homework and follow the advice in this book, and you will be able to avoid such costly mistakes.

Very large weldments are not often thought of in light of welding automation, but some automation is possible in almost any welding situation. These huge weldments include:
- Ships and submarines
- Metal bridges and causeways
- Nuclear pressure vessels, not so much for their size, but their weight
- Mining equipment, such as the huge shovels that can scoop out a load of earth the size of a house
- Buildings and other architectural applications
- Oil rigs and deep sea oil platforms
- Crane booms
- Land-based oil storage tanks and liquid gas containers

Machine welding is the most common type of automation utilized in such applications. This is mainly due to the generally unique joint configuration. For instance, a submarine shell may have no two welds that are exactly the same, yet mechanized welding can greatly speed up a 15-pass weld around its circumference.

Part volume and production rate required

Volumes of one, like the previously mentioned submarine, quickly limit your options as far as determining the type of automation to use. However, what if you produce a large number of commercial air compressors per day? Would a robot be the best solution? Or maybe a cell made up of several dedicated machines? Figure 3-1 shows the compressor which we will be discussing.

If only a single design of compressor exists, and you are doing 200 per day, then some good welding design could limit the amount of welding on such an assembly. With a few welds, and some high, consistent volumes, you might lean toward hard automation for this application, since changeovers will not be expected, and the flexibility of a robot is not necessarily needed.

One machine might weld the caps on the ends of the tube to form the compression tank. This can be accomplished by a single machine with dual torches, where the welding fixture is mounted in a lathe. As the torches begin welding, the lathe begins turning until a complete revolution has occurred, thus completely welding both ends simultaneously.

The next machine might weld the tank to the four mounting brackets. A machine with eight torches can weld all four brackets at the same time. Welding with eight arcs concurrently boosts productivity remarkably when compared to manual welding.

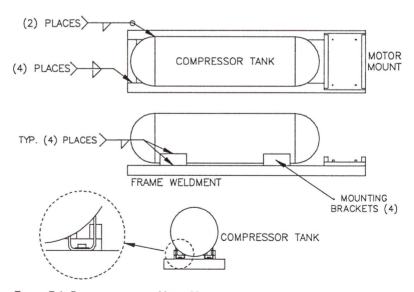

Figure 3-1. Compressor assembly weldment.

The third machine could weld the brackets to a frame, providing a complete product ready for the paint booth. By arranging the machines in a cell, each of the three welding operations will be internal to each other, in contrast to a robot making each weld in series. A robot could possibly be as much as three times slower than the three-machine cell, and the price would not be significantly different.

Now, on the other hand, if there are four significantly different designs of compressor, and the quantities are smaller, say, 20 of each per day, then you will not be able to afford a three-machine cell for each of the four compressor models. This would require 12 machines. There are two choices here. First, a cell of hard-tooled machines might work if they are provided with flexible tooling, allowing any of the four models to be run on each machine or, with quick change tooling, to minimize down time due to changeovers. Changing over four times per day will be quite time consuming without such specific provisions. The second choice is a robotic welding system.

Unlike hard tooled machines, robots can automatically weld whatever part is put before it, providing, of course, that a proper welding program has been created. An effective robot cell may have either four welding stations, each dedicated to a particular compressor model, or flexible tooling like that on a hard tooled machine, which would allow the operator to load any of the four models in any order. The robot, by a number of different means, could be manually or automatically instructed which weldment was in which fixture, so it could call the correct welding program automatically.

Part configuration

The next item of concern is the physical configuration of the parts to be welded. If the part is very large and lopsided, then it might be difficult actually to manipulate the part in order to present the welds to the machine. This may indicate the need for a robotic system, since robots can be taken to the welds when the welds cannot be brought to the robot.

If the parts are small and symmetrical and require only one or two welds, then the machine used to weld them can be quite simple and small. Generally, parts that are small and simple, with only one or two welds, are most effectively automated with hard tooled machines. Even weldments with many welds can be welded with hard automation, by implementing a cell of hard tooled machines, each dedicated to only one or two of the welds. As the part moves from machine to machine, additional welds are applied until all of the welding is done. This allows individual welds actually to be applied internally to each other, since they are done simultaneously on different machines. This method provides much higher rates of production than a single robot can, since a robot will typically have only one arc going at any given time.

If the welds themselves follow simple contours, such as straight lines or true circular arcs, then hard automation is a good solution, but, if the contours are more complex, then flexible automation will be necessary, since simple hard tooled machines are not capable of following more complex paths. Some welds are so complex that robots are the only solution, due to their flexibility and programmability.

Length or volume of welds

If the length of the weld to be made is 60 feet, then, obviously, it is not feasible to put the weldment into a machine and manipulate it during welding. A method is needed to take the arc to the part, wherever it may be. In this case, mechanized welding is a good method of automating the process. It is simple to set up, inexpensive to purchase, and the returns on such an investment are typically very favorable.

If you only weld one of a particular weldment per day, but the total cycle time for each weldment is 8 hours, then automation may still be a good solution. This typically happens when lots of large, multipass welds are to be made. In this case, a large, sophisticated robotic welding cell might be the equipment of choice. The important criterion here is that the parts are repetitive, even if you build only one per day. What allows such an application to benefit from automation is the sheer volume of weld going into the part. The investment in initial programming time can be significant, but such applications often see phenomenal improvements over manual cycle times. Since an automated welding system can reposition the weldment within a few seconds, the cycle times can easily be one-fourth that of manual welding, or less.

If the welds are neither long nor voluminous, but there are a lot of welds on a particular assembly, then a robotic welding system may be the ideal solution. When a part is simply too complex for hard automation, with a lot of welds at all different angles and lengths and sizes, then robots are often the only type of equipment available to complete the job effectively.

Relative positions of welds

The positions of the various welds on a particular assembly has a lot to do with the type of automation one would choose. If all of the welds are parallel to each other and on the same plane—for example, the reinforcing C-channels welded to the side wall of a dump truck body—then more simple, linear travelling automated gantries can be used effectively. Figure 3-2 shows one such machine. This linear gantry welder welds channels A, C, and E at the same time, and then the operator indexes the weldment so the machine can weld channels

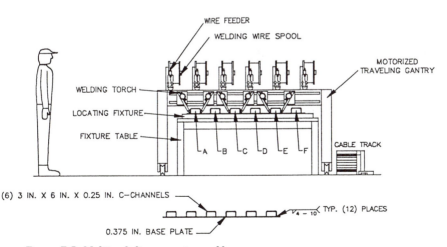

Figure 3-2. Multitorch linear gantry welder.

B, D, and F. Two trips along the floor-mounted rails are all it takes to weld the assembly completely, and the controller can be programmed to create the skip welds automatically that are called out by the weld symbol. These types of machines are especially effective, since several welding torches can be used simultaneously to make parallel welds, and, at a fraction of the cost of a compatible robotic welding cell, the investment is usually much more prudent.

Such a machine might cost from $25,000 to $50,000 or more, but this is much less than a robot cell designed to perform the same task. Since six welding torches are simultaneously depositing weld metal, the productivity of this machine does not even compare to the single arc of a robot, not to mention how it would fare compared to a person.

On the other hand, if the welds occur at all different angles to each other, then, of course, a linear gantry would not be effective. A transfer line may work well in such instances, since the part can be transferred from station to station, each station performing a different weld or set of welds. The drawback with such transfer-line automation is the requirement for high volumes in order to justify the generally large expense. A better solution for small volume applications would be a robotic cell. Although utilizing only a single arc, the robot's flexibility allows it to access welds in many different planes and angles.

Very common applications for multitorch welding include the welding of brackets, which have two parallel welds—one down each side of the bracket; welding of the four corners of a steel box or container simultaneously; and welding a circular weld with two torches—each doing half of the 360 degree circle, in order to cut the arc-on time in half.

Number of weld passes per weld

If your prints call for a 1-in. fillet weld, chances are your automated process will require a dozen or so passes to complete the weld. Not all types of automated machinery are capable of applying multiple passes accurately and quickly, with few defects. Very simple mechanized welding equipment can be used for multiple passes, where each pass is manually offset by the operator.

Other multipass operations, such as cladding the surface of an axle, can be done with little operator intervention, since the machine can be set up to index a previously defined amount after each complete revolution of the axle. After making a complete pass the length of the axle, the machine can automatically return the opposite direction, indexing the welding torch up to offset for the previously applied weld metal, and again indexing sideways with each revolution. This setup allows several passes to be made while the operator is constructively occupied elsewhere.

Robots are especially effective at applying multipass welds, especially around more complex contours. The user must be very careful, since many robots are simply not capable of multiple pass welding, but those that are capable can be quite user-friendly. If your application includes multipass welding, it is wise to request of the vendor a demonstration of their equipment. Be sure you understand how the multiple passes are programmed and how the software actually computes the paths of the various passes. Then you will be able to compare equipment more accurately and will be sure of purchasing the most effective solution for your application.

Weld quality or strength required

If the final product to be welded has very high aesthetic requirements, this may affect the type of equipment you purchase. For instance, if extremely high quality is needed, then you may consider TIG welding. TIG welding provides beautiful, highly consistent welds that will appear virtually flawless. However, certain challenges are present with the TIG process, which are absent when welding with MIG, including the need for very high piece-part accuracy; very high machine accuracy; precise control of the welding arc and filler wire, if used; and additional hardware and software requirements.

On the other hand, if weld quality really is not that important (yes, there may be applications that require that the parts simply stick together), then less expensive welding power sources and less sophisticated fixturing may be acceptable. The smart vendor will provide equipment that represents the best value for the application in question.

Identify the Correct Process

First, identify the welding process to be used. This may not be as easy as it sounds. The process you are now utilizing may not be ideal. Welding wires, gases and equipment have been changing considerably lately, so if you have been doing it the same way for the last 10 years, there may be new developments that can save you money, increase your weld quality, reduce smoke, or improve weldability. You owe it to yourself to determine the very best process available for doing the job at hand. Plenty of information resources are available to help with this task, such as trade shows, welding publications, and assistance from welding suppliers, consultants, and other welding specialists.

A fabricator of heavy machinery welded with dual-shielded flux-cored wire for quite some time. Upon initiation of a robotic welding project, the robotic vendor recommended solid wire rather than flux cored to prevent stopping the robot after every pass for chipping of slag. After extensive testing, the customer determined that solid wire gave them the stringent metallurgical properties that they required. This saved them more than $0.40 per pound on the cost of the welding wire, and they had not yet spent a penny on capital equipment. (They also cut down dramatically on the amount of smoke being generated; flux-cored wire is very messy stuff.) This illustrates the importance of keeping up with welding technology and periodically rethinking the processes and procedures you use in welding every day.

Innovations are constantly being developed in all areas of arc welding. For example, some manufacturers have introduced computer-controlled welding power supplies. The microprocessors in these welding power sources actually strive to supply the exact values at which weld quality is optimized. Some power supplies are designed especially for welding of aluminum. Computer control makes them so easy to set up and weld that it is actually difficult to create a bad weld on aluminum plate.

Welding wire suppliers are continuously improving their products to provide wires that produce less smoke, better penetration, and higher weld quality. The same is true for shielding gas manufacturers, who are always inventing better gas mixes to accomplish various goals related to weld quality and ease of operation. Manufacturers of welding torches, power supplies, bulk wire dispensers, and fume control devices continue to improve their products in this very competitive industry, and the winner is you, the end user. Do some research, and attend welding and fabricating trade shows. These shows can be a valuable source of information as you become more and more educated in the field of welding and automation.

You should know that some processes lend themselves more readily to automation than others. A large majority of the welding robots installed in the U.S. are for MIG welding. This is because applications for MIG welding are simply more numerous and also because the MIG process is much more "forgiving" than TIG welding or plasma arc welding. If your welding operations require automation of a TIG process, then that will immediately limit the field of vendors from which you will choose. Resistance welding is also quite commonly automated, especially in the automotive industry. If resistance welding is the process in question, then many arc-welding equipment suppliers will not even quote such a system.

Identify the Correct Type of Automation

This is a critical step and one that you should not take lightly. Large variations in price can occur among the various styles of automation, which will greatly affect the profitability of the system and its payback period.

First, consider the complexity of the subject weldment. As a rule, the very simple joints, such as circumferential welds for tube-to-flange welding or straight line welding for attaching brackets, are the best candidates for hard automation. If the weldment has 75 welds, all at varying angles and attitudes, then maybe a robot will be better suited to your needs. Remember, however, that a robot will typically have only one arc on at any time. Some complex parts can be welded by more flexible automation, which might include a transfer line, in order to increase the throughput of the system. These line-style systems often cost significantly more than a robotic welding system, but the productivity of such systems can be incredible. The automotive industry uses such technology extensively, due to the tremendous part volumes that they must generate.

Although I have been preaching the ability of robots to weld very small lot sizes efficiently, the part volumes you intend to run do play a significant role in the type of automation that best fits your needs. Small run sizes can normally be better handled by a robotic system, as long as the part runs repeat. There is an initial investment in programming time, but every time thereafter that those parts are run, major savings occur. Even when the parts do not repeat, and you know you will never see them again, there are robots available with the "lead-through teach" teaching mode that allows the user to decrease hard programming time greatly. This also makes robotic welding more accessible and affordable for smaller shops, since such robots are generally less costly than others.

The life of your contract may be a determining factor in which type of automation you purchase. If the product life is very short, then the ability to retool the system quickly and easily is important. Robotic systems are usually easier to retool than other styles of automation, and their programmability gives them maximum flexibility for future use. If the product life is very long, then perhaps a combination of hard automated machines is more cost effective.

The final consideration is your budget. Sometimes, no matter how you slice it, you just might not be able to come up with the money for what you believe is the ideal solution. Do not give up on welding automation. A very simple, inexpensive machine can save you money and pay for itself in a matter of months. There are even robots available now for under $40,000, although you must be careful when making purchases like these. The low prices can lead to lack of services, which are especially important for first-time robot users.

Please refer to Figure 3-3. In case 2, rather than acquiring both machine 1 and machine 2, just installing the first one will save the user a significant sum the first year, maybe enough to pay for the machine while increasing the weld quality and productivity. The user may then buy the second machine a few months later and thus slowly put together a cell that can net big profits the third, fourth, and fifth

---

Here we are comparing the investment vs. productivity for robots and for hard automation on the same welding job. Envision a 5-year contract for welding steel frames for personal computers. The four corners are to be welded, and two identical brackets are to be welded on opposite sides. Case 1 below utilizes a robot to weld all of the welds. Case 2 implements two dedicated machines, the first with four welding torches to weld the four corners simultaneously, the second with two torches to weld the two brackets simultaneously.

Case 1 (Robotics)

Cost for robot system	$100,000
Robot cycle time	69 sec
(60 sec arc-on time plus air cut time)	

Case 2 (Hard Automation)

Cost for two automated machines	$120,000
Hard automation cycle time	
Machine 1	19 sec
Machine 2	15 sec
Effective cycle time	**19 sec**
(Since the machines operate in parallel)	

A 25% increase in cost for two dedicated machines provided more than a 70% increase in productivity.

Figure 3-3. A comparison of investment and productivity for robots and for hard automation on the same welding job.

years of the contract. The customer has spent a little more than the price of one robot cell, but the hard automation will be outperforming the robot three to one, since the cycles are internal to each other.

Look one more time at the above example. A little more was spent on the hard automation for a lot more productivity. The machines are paid for after the first year or two. No programming skills or special knowledge are required as with robotics, thus the learning curve is very short, and most hard automation has very simple controls, fewer parts, and less maintenance than robotic cells. All are great arguments for such automation. On the other hand, after the contract is over, a robot is much more easily retooled for other jobs, even jobs that may be totally unrelated to the computer frames. This is one of the major advantages to robotics, and we will discuss this flexibility in much more detail later.

Integration With Other Plant Processes

It is essential to study the effects of welding automation on your overall production scheme. If welding was once a bottleneck, the automation may eliminate it. *Now* where will the work pile up? If you previously shot blasted your parts after welding before they went to the paint booth, can you now blast them before welding instead? If you do, will you need to blast them again afterward to clean up the mess from the welds? Do the parts have to cool down before painting? Does the operator need to finish the 15% of welds the machine could not reach? Let us take a look at the effects of automated welding on a variety of plant functions and operations.

Welding design

It all starts at the drawing board. The easiest way to deal with welding is to eliminate it. The process of designing weldments is rather unique, and, without an experienced voice around, you may end up with a lot of squabbling between design engineering and manufacturing. Be sure to agree on this, however: welding is a hot, dirty, expensive, labor-intensive process, so to eliminate it is to cut costs, plain and simple. Some good books are available from various sources regarding design of welded structures. Contact the AWS (American Welding Society), or SME (Society of Manufacturing Engineers), or other similar organizations for a publications catalog. These are good sources of information about welding, as well as most other types of manufacturing processes.

The parameters about which you should be concerned include the following:

- Parts should be designed for ease of handling and fixturing. If component parts are ungainly and heavy, hoists will be needed to load and unload them, and if contours are more complex than they should be, the fixturing may be more elaborate than necessary. Design parts to be loaded into and out of fixtures and to be quickly and easily clamped and located.

- Parts should be designed for easy access to the welds. If the parts themselves block access to the welds, then an automated system will be less productive, since the operator will need to do more of the welding. In the same way, if fixturing blocks access to the welds, then the machine cannot be utilized as effectively as it might be. Be sure to leave plenty of clearance for the robotic torch to reach the welds and to do so with the proper torch angles.

- Design the fixture so that parts can be unloaded. This sounds very simplistic and obvious, but it is not funny when it actually happens. Component parts can be easily loaded into a fixture, but it is possible to weld these components into a rigid assembly that cannot be unloaded.

- Design the weldments to minimize the number of components. By doing this, you will hopefully be decreasing the amount of welding necessary, thus eliminating cost, and you will be decreasing the costs associated with a plethora of component parts that are not really necessary.

Manufacturing of components

In chapter 1 we mentioned the importance of repeatable parts for welding automation. The machine will think that the weld joint is in the same place every time. If it is not, your welds will be mispositioned. The fact is, in many cases where automation is installed and welds are being mispositioned, the machine is blamed immediately. This is especially true of robotics, since it is easier to blame a programmable, servo-operated multiaxis robot for making a mistake than to root out the real problems. If the machine is not actually at fault, and the component parts actually have some problems, then that means someone has some work to do to correct the problems with the parts. It is easier just to fault the machine; but no one said that automation is the easiest route to take.

The truth is, such problems are rarely the fault of the machine. Sure, there are occasional problems, but, once a robot is programmed or once a hard-tooled machine is properly set up, things should not change. It is no coincidence how such problems usually come on the heels of a new batch of parts being delivered to the welding cell, a new operator being assigned to the cell, or some other change in the way things are usually done. True, a machine should not be subject to such minor changes as a new operator, but sometimes people try to "fix" or change something for the better, when they really do not understand the process properly to begin with. This will be addressed further when we talk about who should be involved in such projects and who should be running the machine once it is installed.

The reason for all this talk about inconsistent parts is to point out the importance of accurately manufacturing components for automated welding. Even when you think your parts are being manufactured to tight enough tolerances, there can still be problems. A common cause of mispositioned parts is in using different datum points for locating components. If a particular sheet metal component is located from one edge during bending and punching, then the same datum should be used in the welding fixture. If you locate from a different datum point in the welding fixture, then the end of the part at which you want your tolerances to accumulate may float, so that the welding machine can never find the weld joint. Small details like these can drastically affect the performance of a welding cell.

You may have to admit to yourself that you are simply not yet ready for automation. If CNC punches are needed to reach this tolerance goal, then you may be better off starting there. However, you must not use that for an excuse to eliminate the sometimes difficult decision of purchasing a welding machine. The entire sequence of events should be part of a plan to better every one of your manufacturing processes. Long-term planning is something we often do not do enough of. If your goal is truly world class manufacturing, then addressing your welding needs is essential for your success and growth.

Other processes can impact the readiness of your components for automated welding. Brake presses, shears, flame cutters, stamping presses, tube benders, roll formers; these all have certain inconsistencies that can wreak havoc in the welding department. You may be experiencing this to some extent already, even where manual welding is concerned. If your plant is like most in this country, the welders are often given the responsibility of taking up the slack for all the other departments in the company. If gaps between parts are too big, the welders can fix it. If the parts need a little "fine tuning" with the ten pound hammer, then the welders can do it. The welding puddle can hide a multitude of sins, but I guarantee they will come back to haunt you if you try automating. A new attitude must take hold, not just in the welding department, but in all the processes that affect the end product. This is why it is especially important that automated welding projects be embraced by the management, with a commitment to do throughout the plant whatever it takes to make it work.

Postwelding processes

We have been discussing processes that take place before the welding is done, but it is important to consider what happens after welding. Many companies find themselves doing a lot of grinding or other cosmetic work to the welds before they are painted. Weld spatter can accumulate on the parts, the welds may not blend into the base material as smoothly as they should, or maybe the welds are just plain ugly and need to be dressed up. Automation can address these problems very effectively. We will discuss in chapter 4 how weld quality can improve. There are instances where companies have completely eliminated grinding of welds, specifically due to the installation of a welding robot. This can result in very significant labor savings and should be quantified and included in the justification of the equipment.

A common application for robotic welding is fuel and air tanks, which must be leakproof. Very consistent results are being seen in such installations, provided incoming parts are properly manufactured. Besides providing leakproof welds, automated welding systems are very adept at providing welds that just plain look good. Due to the precisely consistent travel speeds and other welding variables, the welds deposited by machines are not only superior in quality to manual welds but also are visually superior. This alone may be enough justification for purchasing a robot, depending on your product line and its aesthetic requirements.

Summary

This has been an informal look at some strategies and reasoning behind the automation of your welding operations. Although subjects were briefly covered here, subsequent chapters will focus more thoroughly on each topic. Here is a summary of chapter 3:

A. Introduction

    1. Automating your welding requires a defined strategy.

B. Identify the correct process

    1. Automation may require you to—or allow you to—use different materials than you are now using. Consult your own in-house expert, or someone from the outside who is qualified to help with such decisions.

    2. Some processes lend themselves more readily to automation.

C. Identify the correct type of automation

    1. Various automation types vary widely in cost.

    2. First, consider the complexity of the weldment. A robot is flexible but more expensive. Hard automation lends itself better to multiple torches.

    3. Programmability of robots makes small batches easier and more profitable to weld.

    4. Longer contracts may lend themselves to hard automation.

D. Integration with other plant processes

    1. If you eliminate the welding bottleneck, are you creating another one elsewhere?

    2. Design your product to eliminate and/or minimize welding.

    3. Inconsistent components are the archenemy of welding automation.

    4. You may need to purchase other machines first in order to build parts consistent enough for automated welding.

    5. Automation can save money by eliminating postwelding processing. Automation can provide highly aesthetic welds.

4. Justifying Your Investment

To face facts, if you cannot prove that welding automation will save your company money, then it probably will not fly. The good news is that it is difficult NOT to show significant savings, especially if you consider the *real* savings potential.

If you use only labor savings for your justification, you are cheating yourself out of a lot of potent ammunition. Many automated systems are justified on labor savings alone, but you need to get a realistic idea of everything that the new technology will do for you. This is the truth; if every plant manager/owner/president/engineer knew how much money he or she was pouring down the drain from inefficient welding operations and the kind of savings available from automating some of these processes, there would not be enough vendors around to supply the equipment. By including the real savings, most of you who read this book and complete the worksheets at the end of this chapter will see paybacks of less than 2 years, and many of you will prove to yourselves that an automated system can pay for itself in 6 to 12 months.

This entire chapter is devoted to showing the reader how welding automation can be one of the best investments you can make in your plant. There is no other way to explain the fact that successful robot users rarely quit after the first one. Those who are doing it right almost always end up with multiple automated systems.

In the pages that follow, I will be describing various sources of savings, so you can calculate a figure for return on investment (ROI) or payback if you so desire. When you calculate your ROI, remember to use all the sources of savings in order to give a clear picture of the true savings available to you. For now, just read on and absorb the concepts that are focal to understanding how automation can change your business.

Labor Savings

Labor savings usually provide the biggest numbers when calculating payback. Although some companies approach automation with the goal of eliminating people, the reverse usually becomes true; more people are hired to keep up with the increased work load provided by new business. Then follows more automation, then more people, and a thing we like to call "growth." What this all means is that while you may not actually decrease your number of employees, you can be assured that automation will provide higher productivity when compared to manual welding. You may be asking, "How much higher?" Let us explore the possibilities.

We have already learned that a typical robot cell can accomplish the same amount of work as about three to four workers on the average. This alone can provide an attractive payback. If your shop rate equals $8.00 per hour, then 2000 hours per year comes to $16,000.00. Benefits and other costs associated with hiring and keeping a good welder may multiply this number by 1.5, giving a yearly cost of $24,000.00 per person. If a robot can indeed displace four workers, then the yearly savings comes to (3) ($24,000.00) (you still need one person to run the robot) or $72,000.00 per year. With a two-year payback scheme, you can justify a $144,000.00 robot system. Now, for some of you, this amount could purchase a lot of robot. The investment can be better than most people realize.

Internal load/unload

Just where does this labor savings come from? The first giant stride forward comes from the ability to load and unload during the welding cycle. Obviously, this is not necessarily true with all types of automation, such as certain types of smaller, less expensive hard automation. However, with this type of machine, the cost is usually less, and the welding process is so fast and efficient, that paybacks

---

The current manual welding sequence for backhoe buckets is as follows:

1. Load parts into tack fixture and tack	10 min
2. Weld as much as possible	50 min
3. Unload part, finish welding, send it away	15 min
4. Repeat process	
Manual cycle time	75 min

Effective Cycle Time: **75 min**

Yearly Volume: **20,000 buckets**

# of Welders Needed **12.5**

(75 min x 20,000) ÷ (60 min) ÷ (2,000 hr/yr) = 12.5

Figure 4-1. Current manual welding sequence for backhoe buckets.

are still very attractive. This ability is most common in robotic welding systems and can automatically double the production rate of a manual welder.

An example that we can use throughout this chapter will be beneficial. Consider the welding of backhoe buckets, as described in Figure 4-1. It takes 10 minutes to load the fixture and perform a few tack welds. It takes the operator 50 minutes to weld as much of the part as the fixture will allow. He then must do 15 minutes of finish welding after removing the part from the fixture. The total cycle time is 75 minutes, since all activities are done in series.

Now, if a two-station robot system is used in our example, as shown in Figure 4-2, then the tacking and load/unload time can be made internal to the robot weld cycle. The robot will be welding buckets at station one while the operator loads/tacks/unloads at station two. Since the operator has 25 minutes of work to do (assuming he still tacks and does the finish welding) then his time is completely internal to the robot's. Already we have reduced the cycle time by 25 minutes, for a savings of 33%, even if the robot still takes 50 minutes to do its welding. See Figure 4-3.

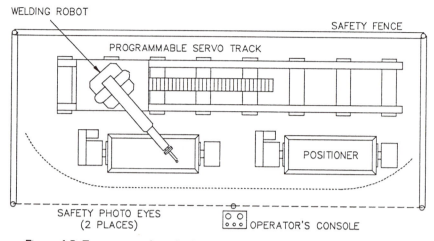

Figure 4-2. Two-station robotic bucket welding cell.

Assuming that the operator's time is completely internal to the robotic welding cycle, and the robot cycle time is still 50 minutes, then the effective cycle time is 50 minutes:

1. Load parts and tack (internal to robot)   10 min
2. Unload, finish welding (internal to robot)   15 min
3. Robot weld cycle   50 min
4. Repeat process
    Manual cycle time   25 min
    Robot cycle time   50 min

Effective Weld Cycle: 50 min

Yearly Volume: 20,000 buckets

# of Welders Needed: 8.3

(50 min x 20,000) ÷ (60 min) ÷ (2,000 hr/yr) = 8.3

Figure 4-3. Effective cycle time of 50 minutes.

Increased duty cycle

The next production gain comes from the ability of the robot to complete the weld cycle faster. Remember our discussion about duty cycle? By moving at speeds of 6 to 10 feet per second, the robot can provide arc-on percentages of up to three times higher than a manual welder. A robot does not have to lay down its torch to flip over a part, assuming a positioner is part of the cell; it does not have to stop to look where the next weld is; it does not go on breaks, smoke cigarettes, or drink coffee.

In our example, we will assume that the actual robotic welding cycle is a conservative 18 minutes, or 32 minutes faster than the welder. Since the operator still needs 25 minutes to do his work, the effective floor-to-floor cycle time is 25 minutes, and the operator is now the limiting factor. This is a savings of 67% compared to the original 75 minutes of manual welding. See Figure 4-4.

The robot can actually weld in 18 minutes what the operator welded in 50 minutes, yet the operator still requires 25 minutes for his work, so the effective cycle is now 25 minutes:

1. Load parts and tack	10 min
2. Unload, finish welding	15 min
3. Robot weld cycle	18 min
4. Repeat process	
Manual cycle time	25 min
Robot cycle time	18 min
Effective Weld Cycle:	**25 min**
Yearly Volume:	**20,000 buckets**
# of Welders Needed:	**4.2**

(18 min x 20,000) ÷ (60 min) ÷ (2,000 hr/yr) = 4.2

Figure 4-4. Effective cycle time of 25 minutes.

Now, we do not want the robot to be waiting for the operator. So some of the 15 minutes of finish welding is given to the robot to balance the workloads. Notice that we do not give all the welding to the robot. Sometimes it is better to share the workload than to have the operator standing around for 10 minutes every cycle with nothing to do. If the operator's work is diminished to 20 minutes, and the additional work given to the robot increases its cycle from 18 to 20 minutes, then the effective cycle time is 20 minutes. This is 73% faster than the manual process, the operator is less fatigued, and the weld quality is probably better. The final result of our robotic welding installation in the bucket cell appears in Figure 4-5.

I want to emphasize the value in sharing the work between welder and robot. It does not always work this way, but often the productivity of the cell can improve if the robot does not do all the work. Remember that the operator is doing about 20 minutes of work. If the robot was to do all the welding, the first consequence is the increased expense of the system. Dual-axis positioners may be needed for extra flexibility, which could easily add $30,000 to $40,0000 or more to

---

The final result, after splitting the work between the robot and the welder, yields an effective cycle time of 20 minutes:

1. Operator's work (internal to robot cycle)	20 min
2. Robot weld cycle	20 min
Total Cycle Time:	20 min
Yearly Volume:	20,000 buckets
# of Welders Needed	Total of 3.3 Welders Needed

Compare this to the numbers for manual welding: cycle time of 75 minutes, requiring 12 welders.

Figure 4-5. Results of splitting work between the robot and the welder.

the price of the system. The second penalty is decreased production. If the robot does all the welding, then the cycle time may be as high as 30 minutes. This is an increase of 10 minutes, or 50%, in the achievable cycle time that we already calculated. By sharing the work, the operator stays busy, the robot never has to wait for the operator to finish loading or unloading, and the overall productivity of the cell is much higher. These numbers are based on real cases. Companies commonly see very similar results, often justifying even quite expensive systems (over $400,000) in just 18 months.

Reduction or elimination of changeovers

The next "quantifiable" savings come from the reduction—or sometimes the elimination—of changeovers. Some robots now have a background editing feature, which allows the user to set up the next program while the robot is actively working. Even without this feature, if a robot has four stations, as shown in Figure 4-6, then the robot can be working alternately between stations 3 and 4, for example, while someone else changes the fixtures on stations 1 and 2 to a different part number. There are actually two people at the cell

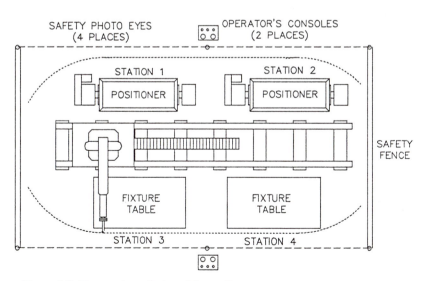

Figure 4-6. Four-station robotic welding cell.

temporarily—one operator and the technician making the fixture change—but changeovers are typically the responsibility of the welder, who cannot be making parts if he is changing fixtures. In the above scenario, production parts are pouring from the robot cell even while the changeover is being accomplished.

Some systems feature completely automated changeovers. With "smart" fixtures that are equipped with servo mechanisms, the fixture details can move, as prompted by a programmable controller, to accommodate different part configurations. These activities can all be launched via the robot controller, based on some input that informs the robot which part is being presented to it. Therefore, the downloading of the program, the readjustment of the fixtures, the welding of the parts, and the indexing of the fixtures into and out of the cell are all automated. This technology, of course, is costly, and not very common. Such sophistication can usually be justified only in special applications.

The reduction of changeover time is especially beneficial to the job shops that normally work with small lot sizes and have frequent changeovers. Using our earlier example of welding backhoe buckets, let us assume that it takes 45 minutes to change over to another type of bucket. Since this company produces a lot of aftermarket product, changeovers happen about three times per shift on a particular line. A total of 135 minutes is spent per shift, or 270 minutes per two-shift day, just for this unproductive task. By installing a multistation robotic welding system, the changes take less than 5 minutes each. This saves 240 minutes, or 4 man-hours, per day, or 960 man-hours per year. Remember this last number for later.

Reduced labor on subsequent operations

Now, these buckets need to be painted before they are shipped. However, a little clean-up is necessary first. Some weld spatter needs to be ground off, and some of the excess weld metal needs to be smoothed out a little. Currently, the operators are spending about 15 minutes per bucket for this purpose. By providing more consistent quality, the robot eliminates almost all of the spatter, and weld size is consistent and uniform. The cleanup is reduced to 5 minutes per bucket, for a total labor reduction of 3,333 man hours per year. Remember this number, too. We will use it shortly.

Other savings

There are more ways in which automation can save you money, but we will not discuss them in great detail. These may include reduced or eliminated inspection costs, due to the higher quality and more predictable performance of the automated process. By eliminating inspection, which is often performed by the welders themselves, you will cut costs and improve lead times for your customers. However, you must be careful that the decision to do away with quality checks and inspections is preceded by a detailed examination that proves that the process itself is reliable enough. To end inspections prematurely before your process is in control could spell disaster.

Other possible savings include the reduced costs incurred as a result of field failures or warranty costs. By providing higher quality welds, weld-related defects are reduced in number and severity. This leads to fewer field problems and, therefore, lower after-sale service costs.

Training time and learning curves can be reduced. Many companies are having trouble finding accomplished welders, so they are hiring nonwelders, then training them in-house. Although I highly recommend that a skilled welder be responsible to run a welding machine, at the same time I must concur that you may have no skilled welders available. If this is the case, then it is easier to teach a nonwelder how to run a machine than to teach him or her all the intricacies of welding. Your own experience and creativity are the only limits to discovering other legitimate savings like these.

Where the rubber meets the road

There are worksheets at the end of this chapter for calculating a rough cycle time for robotically welding one of your assemblies. Follow the directions carefully. Ask an experienced welder to help. The numbers you end up with may or may not be very accurate. You must remember that, more so than most manufacturing processes, welding is very unique and the application of automation quite specialized. It is mandatory that you enlist the help of someone with experience in order to answer all the questions effectively and to start you down the path to automation success properly. However, these worksheets will at least provide you with a feeling for what you can accomplish with some automation in your welding department.

Material Savings

People overweld, and that costs a lot of money. Figure 4-7 shows how the volume of a fillet joint disproportionately increases with small increases in the fillet size. For instance, a $1/4$ in. fillet joint has a cross-sectional area of 0.03125 in.^{2} By increasing the fillet size by only 25% to $5/16$ in., the area increases to 0.04883 in.^{2}, requiring 56% more weld metal. If your welders happen to be overwelding by $1/8$ in., a very easy and common mistake to make, the increase in volume is a whopping 125%. That can add up in a hurry.

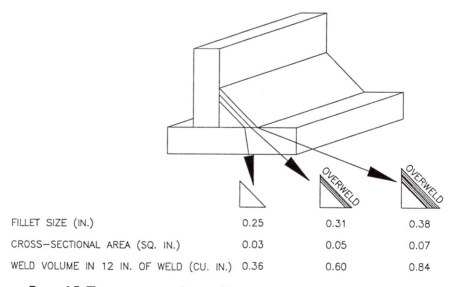

FILLET SIZE (IN.)	0.25	0.31	0.38
CROSS-SECTIONAL AREA (SQ. IN.)	0.03	0.05	0.07
WELD VOLUME IN 12 IN. OF WELD (CU. IN.)	0.36	0.60	0.84

Figure 4-7. The consequences of overwelding.

The backhoe bucket manufacturer uses about 5 lb of weld metal per bucket while manually welding. If automating can save even a conservative 25% of this by eliminating overwelding, then at $0.60/lb, the annual savings can exceed (20,000 buckets) × (5 lb/bucket) × (0.25) × ($0.60/lb) = $15,000.00/yr. Add to this the savings of using bulk wire and gas (if you are not already doing so; you do not need automation to utilize bulk consumables).

Increased Production Capacity

There is a real cost involved in dedicating floor space and production time to a particular product. How you identify and quantify these variables is probably unique to your company. But quantify we must.

First of all, automation can make your resources available for other productive work. I will say it again; very often, automation does not cause layoffs, but increased production capacities help you gain more work, and the end result is growth, new projects, an increased customer base, and another step toward world class manufacturing.

How should you quantify these values? Consider this: if you are a job shop, and the presence of an automated welding system helps you land a contract you otherwise would not have, then it is easy to determine the worth of the system. This alone can justify complete robotic welding systems, and often that is exactly the way it works; a company decides they need the robot for a particular job, and, without it, they lose the contract. It is an easy choice to make. Two-year paybacks and other such nonsense become meaningless in light of such opportunities. The robot is purchased because it is needed, not because a bunch of number-crunching provided the correct figures for the board members.

Because automated welding systems can accomplish some factor above and beyond what a manual welder can, they often consume less floor space per end-product produced than the equivalent number of manual welding stations. For some with plenty of room this does not matter. For others, floor space is like gold. Weigh the importance of this factor yourself, and determine the value of producing the same volume of parts in only a fraction of the space.

World-Class Market Response

The ability to produce small lot sizes economically can have a tremendous effect on your bottom line. One facility I recently visited has only two of any given subassembly in the plant at any time. When component parts get low, an order is faxed to their supplier, and the parts arrive within a couple of days. By welding the components together in lot sizes of one, this customer can experience true JIT delivery, not only to their own internal customers (in this case, the assembly line), but to their outside customers, the ones buying the end product.

Amazing things can happen when this level of efficiency is attained; you can build a product according to your distributors' or customers' orders; there is virtually no inventory, so there is no inventory cost; and aftermarket and replacement parts can be built simultaneously with OEM parts with no disturbance in product flow and with no additional setup. The advantages are limitless.

Just as you demand better delivery and higher quality from your vendors, so will your customers demand it of you. Is your welding department a bottleneck? Do you have more employees doing welding than you would like? Do you get complaints from the welders about "too much smoke, too hot, too dirty, too hard to do my job correctly?" Automation can solve all these problems, in addition to providing higher quality and reducing your cycle time significantly, all while saving your company money.

You may be asking yourself why this pep talk is included in a section about justifying the cost of automation. It is indeed difficult to assign a value to some of the advantages just described. Now we are down to the philosophy and risk underlying welding automation. Yes, there is a risk. The risk is in doing a partial job when implementing the solutions, and not understanding what to expect from automation. This book will eliminate most of the risk, so that you can make an informed decision about how to automate. The philosophy is entailed in this question: Do you want to remain competitive? Most other plant processes have become automated. Most companies are very competitive in machining, computerized bending, punching, and stamping. Where will manufacturers get the edge they need? By paying attention to the one process that has been neglected the most: welding!

Intangibles

Just a few more comments to close out this section on justifying automation. A few other hard-to-quantify benefits should be discussed here briefly, the first being safety. Robot operators breathe fewer fumes, are exposed to fewer sparks and less molten metal down their pants, do less man-handling of heavy parts during welding, and, in general, have an improved outlook on their jobs. This alone can be assigned a value by assuming that, above and beyond the additional productivity of the automated system, that the workers' levels of performance and productivity will increase. This is a real savings, but unfortunately is rarely used in proautomation arguments. Improved safety also contributes to fewer job-related injuries, which can cost some companies tremendous amounts. One back injury can pay for an entire automated welding system.

It All Adds Up

We could continue to boil down the advantages to ever diminishing degrees, but we should have enough to go on already.

Let's get back to our bucket example, and see how we came out.

1. The decreased cycle time has saved 55 minutes per bucket. For the annual volume of 20,000 buckets, that comes to 18,333 man-hours saved. By using our $12.00/hr figure for labor and benefits, the labor savings comes to $220,000/yr.

2. Eliminating some of the changeover time saved 960 man-hours. At $12.00 per hour, this equals $11,520/yr.

3. Reduced grinding provided 3,333 man-hours of savings, for a total annual savings of $40,000.

4. Increased material efficiency by eliminating overwelding yields savings of $15,000/yr.

We will not use any of the other factors in our calculations, simply because we do not have to. Items 1 through 4 provide a savings of more than $286,000/yr. A very capable bucket-welding cell can easily be had for less than this. Most companies can appreciate paybacks like this. The good news is, 1-year paybacks are not uncommon in the welding industry. Being a young industry, and usually having the distinction of being the only process left that is done the same way today as it was 20 years ago, welding is a field ripe for the harvest that automation can provide and is providing for many of your competitors today.

You can calculate your own rough cycle times by using the worksheet found at the end of this chapter. Experienced applications engineers should be the ones to do the real cycle time analysis, since there may be some portion of the welding that the machine simply cannot reach. It takes some automation experience to determine that. However, the results from the worksheet should at least give you a feel for what kind of savings you can generate by automating what is probably the most expensive process in your plant.

Summary

It is difficult to prepare a poor ROI for automated welding. If you play your cards right, you will see that it can be a very lucrative investment. Welding is labor intensive, rates are high, and the work is hot and dirty. Automation addresses all of these issues and more, providing a very attractive payback.

A. Labor savings

  1. Faster cycle times are possible.

  2. Automation provides higher duty cycles than manual welding.

  3. Labor for changeovers can be reduced or eliminated.

  4. Labor for subsequent operations, such as cleaning, can be reduced or eliminated.

5. Other savings are available, such as inspection costs, more predictable performance and material usage, and fewer field failures and warranty claims due to higher quality.

B. Material savings

　　1. Significant time and material savings can be realized by eliminating overwelding.

C. Increased production capacity

　　1. Productivity per floor area is increased.

　　2. Other resources are made available for productive work.

D. World class market response is more achievable

　　1. To produce small lot sizes economically increases efficiency, eliminates costly inventory, and saves time and money.

　　2. It is easier to build per the customer's order.

E. Intangibles

　　1. Define the intangible benefits and quantify them!

F. Do the worksheets and see for yourself how welding automation can benefit you

Estimated Robotic Cycle Time Worksheet

Step 1: Calculate Arc-on Time

Weld Size	# of Welds	Total Length (in.)	Weld Speed (ips)	Arc-on Time (sec)
1/8 in. (3 mm)	_____	_____	35	_____ (sec)
3/16 in. (5 mm)	_____	_____	30	_____ (sec)
1/4 in. (6 mm)	_____	_____	18	_____ (sec)
5/16 in. (8 mm)	_____	_____	11	_____ (sec)
3/8 in. (9.5 mm)	_____	_____	8	_____ (sec)
1/2 in. (13 mm)	_____	_____	4	_____ (sec)

Formula: [(Total length) ÷ (Weld speed)] X 60 = (Arc-on time)

Step 2: Calculate Air Cut Time

(# of Welds) X	(Time per move)	=	(Air cut time)
_____ X	(3 sec)	=	_____ (sec)

Step 3: Calculate Manipulation Time

(# of positioner moves) X	(Time per move)	=	(Manip. time)
_____ X	(10 sec)	=	_____ (sec)

Step 4: Track/Shuttle Time

(# of track moves) X	(Time per move)	=	(Manip. time)
_____ X	(10 sec)	=	_____ (sec)

Step 5: Total Robotic Cycle

Add right-hand (sec) column. Divide result by 60. This is your estimated cycle time in minutes. This assumes a 100% duty cycle. For a typical production capacity, divide by 0.85. The result will be a cycle time estimate at 85% efficiency, a standard for the industry.

Total = _____ (sec at 100%)

÷ 60 = _____ (min at 100%)

÷ 0.85 = _____ (min at 85%)

Directions for Cycle Time Worksheet

Step 1: Calculate Arc-on Time

Calculate the total length of weld for each weld fillet size. Enter the lengths, as well as the total number of welds of each size, in the appropriate spaces on the worksheet. For instance, if the weldment in question has six 1/4 in. fillet welds, each 4 in. long, you would enter *24 in.* in the 1/4 in. row under "Total Length" and *6* under "# of welds". Do this *for each weld size,* since larger welds require slower travel speeds.

Calculate the arc-on time by using the following formula:

[(Total length) ÷ (Weld speed)] X 60 = (Arc-on time)

Step 2: Calculate Air Cut Time

Air cut time is the time needed for the welding torch to be moved from weld to weld throughout the welding program. Calculate by counting the total number of welds, of all sizes. Add 1 to this number (since the torch must move home after the last weld), and multiply the result by 3 sec. Three seconds per move is an estimate and may be longer or shorter depending on the average length of move.

Step 3: Calculate Manipulation Time

This calculation is only necessary if a workpiece manipulator is part of the cell. Estimate the number of times the positioner must position the part in order for the robot to reach all of the welds. Multiply the result by 10 seconds. Ten seconds per move is an estimate and may be longer or shorter depending on the average length and complexity of the move and the type of positioner in question.

Step 4: Track/Shuttle Time

This calculation is only necessary if a robot track or shuttle is part of the cell. Estimate the number of times the track must move in order for the robot to reach all of the welds. Multiply the result by 10 seconds. Ten seconds per move is an estimate and may be longer or shorter depending on the average length of the move and whether the robot is mounted to programmable track or an indexing shuttle.

Step 5: Total Robotic Cycle

Add the right-hand "seconds" column for a total cycle time in seconds. Divide by 60 for minutes. This will give you a cycle time estimate at 100% duty cycle. Since even robotic cells are subject to downtime (out of parts, operator on break, changing wire or gas, etc.) it is typical to divide the cycle time by 0.85, which yields a cycle time based on a duty cycle of 85%. This is the production capacity that can be expected from a typical robotic welding cell.

Remember, this is a *very rough* cycle time estimate. Accurate estimates depend on a large number of factors, including: how many of the welds can the robot actually reach?; how much part manipulation is necessary?; will some torch moves take longer than others?; quantity of multiple pass welds; number of touch-sensing routines present; the list goes on. This illustrates the necessity for an experienced engineer to prepare the estimates.

5. Laying the Foundation

If you are like most welding manufacturers out there, you will need to do some homework before starting down the road to automation. It is important to prepare properly, and this task may take anywhere from a few weeks to many months, depending on the current state of your welding operations and other production processes. This is where a welding automation expert will earn his keep very quickly, but there are also a few things that only you can take care of. These will become evident as we dive into the topic.

This chapter begins a section focusing specifically on how to begin the automation process and how to minimize the risks and maximize the profitability of the system that you install. We start here with the task of laying the foundation, without which your efforts could crumble like the biblical house built on sand. The major issues are company vision and the involvement of all plant functions from managers to factory floor weldors. Proper identification of what to automate, how to automate, and why, are also addressed.

Vision

Whether you are an engineer trying to initiate projects that will save your company money, a manager wanting to become more competitive, or a company owner hoping to guarantee that you survive and thrive in the years to come, or a trade welder who takes pride in your work and wants to find a way to better utilize your welding skills and knowledge, welding automation should become part of your vision. I am not just talking about wishful thinking, but you must be convinced in your own mind that automation is key to success and

97

security in a market that we all know is subject to global competition, very cheap overseas labor, and other countries that are already very highly automated. Without this vision, the whole automated welding exercise is futile. Why? Because there are already too many people out there who do not believe that robots work, who think they cannot afford automation, or who are afraid of what the union might think. (By the way, some of the most efficient robotic installations are in union plants, and most are very successful. Trade unions are realizing the value and necessity of automated welding.) Without a vision, you will not stand up under the scrutiny of those around you who doubt. However, with a goal in mind, and being convinced of the end results, you can be successful, probably beyond what you even expect, regardless of what your inexperienced and pessimistic peers in industry tell you. Listen to successful robot users instead.

This book is not a motivational book, but I am interested in seeing industry embrace automated welding. The alternative is to knuckle under to the other manufacturing superpowers in the world, who are even now installing many more welding systems than U.S. manufacturers are. If you are interested in growth, get a vision. Start saving money today, and become a world-class welding shop.

A small job shop I once visited decided to automate. They employed a mere dozen people and had but one consistent product line—automotive door handles. Of course, many other jobs found their way through the doors of this company. Where most companies would have been afraid, however, the competitive owners forged ahead and purchased a small robotic welding system. After the robot had been in use for a few months, I discussed their second robotic welding system with them. I recommended a larger robot, perhaps on a track, that would allow them to pursue jobs involving larger weldments. Though they liked the idea, they decided that a larger robot was not in their immediate future. "There are so many jobs out there for our existing robot, we cannot fill the need. We need another small system. We do not even have the time to think about welding larger parts."

This small entrepreneurship could not keep up with the demand for their robot. Before the first one was paid for, they began discussing robot number two. Of course, it will not be like this with every company, but, rather than viewing automation as a replacement for human workers, you must expect to grow. This is especially true while welding automation—and especially robots—are still coming of age, and not many of your competitors have the automation advantage. Your customers will prefer buying parts from a vendor who will provide the higher quality and consistency that automation provides.

Operator Involvement

Welders are proud of the work they do, and they should be. There is art, as well as science, involved in laying down a good bead, and it is something that takes practice and patience to learn. Your welders have a lot of knowledge in their heads about how to make a good weld, what variables affect the quality of a weld, and how the product flows through the shop; in other words, they are the people that really know what is going on. This is precisely why the welders need to become the automated machine operators. It is difficult to teach a programming wizard the intricacies of welding. It is quite simple, however, to teach a welder how to transfer his skills and knowledge to a robot or other automated machine.

In addition to being the only logical choice for machine operator, your welders also need to be part of the decision making that takes place throughout this whole process of preparing for, specifying, purchasing, installing, and starting up an automated welding system. Let's face it, welding vendors are generally very good at what they do, but you and your people are the experts when it comes to manufacturing your parts. Vendors might unwittingly give you some suggestions that your welders know just will not work. It is much better to flush these questions out right away than to deal with them after the system is built.

Keep no secrets from the operators. They should see the quotations, they should be privy to the savings and payback calculations and should know pretty much everything that the engineers and management know about the project. The welders will recognize flaws in the proposed system configuration and can probably give some very valuable input on engineering features or changes that can make the system more productive, more user friendly, or more profitable.

For instance, what if, during the system runoff at the vendor's facility, the welds develop slight porosity or undercut? What if the throat or the legs of the fillet are too small, too convex, or too concave? Is penetration adequate? Is the part being welded at the correct angle and with the correct welding parameters? Do you know personally what all these things mean? Can you personally spot all of these problems as well as your welders can? Chances are, if you have not had much exposure to welding, then you will not really have a clue what to look for during the system runoff or even if welds that are being created are good or bad. In much the same way, if anyone but an experienced welder is running the machine, then he or she may not know whether the equipment is performing properly or not.

Management Commitment

By now, discussions concerning management commitment are probably redundant. But I cannot stress enough how important upper level commitment is to a project like this. This type of support comes in many shapes, but the lack of it is immediately evident. Be sure that you are not substituting interest and curiosity for true commitment. Commitment requires effort.

A manufacturer with several plants nationwide was a very successful user of welding automation, but when a particular plant installed their first robotic welding system, something went awry. As was the custom in this plant, workers were shifted from job to job every few weeks, presumably to avoid boredom, to help them understand other plant processes, and learn how all the operations fit together. While in some ways, this is not a bad idea, this practice sealed the fate of that robotic welding system. Management had not committed to structuring a classification for operators whose sole responsibility was to run the robot cell. Unlike a manual welding machine, an unfamiliar operator cannot be expected to waltz in and be a robot expert overnight. Why do you suppose robotic welding systems come with a week of intensive training?

The results were predictable. The robot was covered with plastic, and the fixtures were used for manual welding. A *very expensive* manual welding booth to be sure.

It should be no secret to you that the welders who manufacture your product every day are the ones who really know the product. If you keep up with current trends, you have seen that the companies that have given decision making authority to factory floor workers are the ones who are successful. Listen to your welders' suggestions, then have the guts to implement them. This is what commitment is all about. I read somewhere that there are very few million-dollar solutions, but there are millions of one-dollar solutions. Attention to these details can make the difference between a mediocre robot installation and one that sings in four-part harmony.

Again, management commitment can take many forms, but the same rules of thumb apply here as to starting any venture. Do not quit. Grin and bear it during the learning curve. Seek the advice of seasoned robot users. Work with your vendors; do not treat them as antagonists. Utilize the skills of the vendor. Be ready for change, which is something totally against human nature. Expect great things.

Organizing the Team

Who should be involved? Obviously, when it comes time to visit the vendor's factory, or a user's facility, you don't want to pay airfare for 25 people. However, you should take everyone on your team. If you assemble the right combination of minds and personalities, you may be surprised how well the whole project can go.

Management should be involved to lend guidance and support to the team's actions. Management commitment is important for the survival of the project, since the success of welding automation projects always depends on support from the top down. Without adequate commitment at the top, proper resources will simply not be provided, and the project will be doomed to fail.

An engineer who designs your weldments needs to give input. This design function is very critical, especially at the beginning of the project. The parts to be welded must actually be "weldable" by a machine, or the entire automation exercise is futile. Ideally, parts should be designed for automation, and the input of the welders and all other team members is important at this juncture.

A manufacturing engineer or manufacturing foreman who knows how things on the floor go together should be involved. This person will be heavily involved in part design, as well as how the welding system will facilitate parts flow, how the machine and man will interface with each other, and how the addition of automation will affect the daily operations of other plant processes.

The fourth member should be involved in quality assurance (QA), and should have an active role in deciding what is a good weld and what is a bad one. In addition, the QA team member should view the project from a total quality standpoint: How can the automated system improve the weld quality? How will it improve the overall quality of the final product? How will this project enable us to meet our customers' needs better, while decreasing defects and reducing costs? What data-gathering devices should be included to monitor the process? Should we monitor the process itself, or should we inspect parts after welding?

Certainly a welder (preferably two or three), should be involved in the team. These are perhaps the most important team members. Those who are in daily contact with the existing process will know exactly what it takes to automate it. They will pick up on the little details that you and even the vendors may miss, and they will have valid ideas on how to optimize the effectiveness and function of the system in your particular work atmosphere.

If you think it is necessary, a financial representative should be involved. This will help keep the project on track and usually prevents the purchase of unnecessary—and often expensive—options. Although financial expertise is necessary to determine whether the investment is expedient, you should not necessarily buy the system you can best afford. Buy the one that will do the best job for you. Time and again I have seen the costly results of price buying, and it can leave a very sour taste for automated welding.

The team should be involved in every decision. Take them places. Show them robots in action around the country. Go to trade shows, and ask lots of questions. Six heads are better than one. The more you know about what robots can and cannot do, about what is being done successfully and what does not work, the better prepared you will be for your installation.

You have vision, you have involved all the right people, a commitment has been made, and the team is assembled. All that is really left before you dig in is to know where to start. The next section focuses specifically on how to go about initiating a project and in what order the crucial steps should be taken. Since no two companies operate in the same manner, you should adapt this approach to your way of doing things.

Getting the Ball Rolling

This is the toughest part. Set aside a time and place for regular meetings. Do not let constant interruptions dilute the effectiveness of the proceedings. This comes back to management commitment. Whether or not you care to admit it, the plant will survive for a half hour without you. Use the rest of this chapter as your guide, and execute the process step by step, and before you know it, you will be back at square one ready to install your next profitable system.

Step 1: Identify the Project

Chances are, you took this step long ago. Before you invested in this book, you had a desire or a goal to accomplish, and the odds are good that the application you are thinking of is a good candidate for automated welding.

Chapter 2 discussed the various types of automation available, and how they function. Chapter 3 emphasized specifying the correct welding process. We will now deal with how to recognize automation projects, based on our knowledge of automation types and welding processes discussed in these earlier chapters.

The basics

Do you produce the parts on a consistent basis?

If the answer to this question is "yes," then you are on the right track. Daily, weekly, or once a month is no problem. Even if you build a particular assembly only once a year, it can still pay. Those applications that repeat and have become a familiar part of your product mix are very visible to all in the corporation. Successful automation projects that utilize your more popular product lines can pave the way for those later welding systems that may be more challenging to justify, yet are just as valid where saving money is concerned. Of course, the most obvious reason for choosing a repetitive job to automate is that those jobs require more labor and are therefore easier to justify.

Even if the answer to this first question is "no," technologies such as lead-through teaching can be feasible for job shops that do custom, small-batch welding jobs. Lead-through teaching is a technology whereby teaching robot programs is greatly simplified, so the operators' learning curve is thus dramatically shortened. In lead-through teaching, the operator simply grabs the robot arm and leads it to the desired point. A button is pushed that stores the position, then the robot is moved to the next point, and so on. Very little training is necessary, and the time spent teaching the robot becomes a much smaller percentage of the total setup time as compared to con-ventional robots. This allows the user to program a unique part in a very short time, then commence welding. Even if that particular part never comes through your shop again, you have profitably automated its welding.

What are the part volumes?

Volumes do not need to be very high to justify automated welding. This question is mostly required to determine what type of automation is needed. Very high volumes almost guarantee excellent paybacks regardless of the type of automation employed, although the next paragraph describes why high volumes are not necessarily the only key to making automation pay.

In general, at least when dealing with simple weldments, higher part volumes require less flexible machinery. If the equipment you purchase is expected to weld the same parts for several years, then the flexibility of an expensive robotic welding system may not be needed. If part volumes are small, and changeovers are frequent, then a robot may be absolutely necessary.

How much weld volume per assembly?

If you only build four parts a day, but each part has 400 welds totaling 900 inches of linear weld, then it could be a great robot application. If the parts have only two welds, each 3 in. long, but you need to build 500 per day, then automation is still the answer, although a robot probably would not be the best solution. Small, simple weldments can be welded by any number of machines, but, as parts become more complex, and the volume of weld per part increases, then more specific and more flexible automation is required. This is especially true where multiple passes are required or when such multiple passes occur in complex contours.

Is there currently a safety risk for your operators?

This alone is sometimes the only reason needed to purchase such machinery. Safety from fumes is easily addressed by removing the operator from close proximity to the welding arc. By keeping the operator outside the cell loading and unloading parts, the fumes that are generated can be separated by barriers and removed or filtered without the operator ever coming in contact with the harmful ingredients of most welding smoke. Safety from weld spatter and molten metal are also effectively addressed for these same reasons.

If the required welding requires very high amperages, then the exposure of the operator to heat is also removed. Machines rarely care whether they are welding at 200 amps or at 600 amps. If your operators are turning down the welding parameters simply due to excessive heat, then automation can greatly improve productivity simply by allowing use of the optimum welding variables.

Call an expert

For help in identifying good automated welding projects, call on the automation vendors, or an automation expert. These are the people that have been doing it for years, have traveled to various and sundry manufacturing facilities, have learned by experience what works and what does not, and what will be the best solution for you. However, you can and should learn what a basic candidate for automation might look like, so you can have as much input into the process as possible. There is no practical reason why you should not be involved, and every reason for you to be very heavily involved at this stage of the game. Do not let vendors simply sell you what they have. If it is not the right solution, then keep looking until you find someone to solve your problem *correctly*.

Rules of thumb

In general, these are some rules of thumb, one or more of which may indicate that you have a product that is ripe for automation:
- The parts are produced on a regular basis
- You have very high volumes of welded parts
- You have parts with very large amounts of welding required per part (may be high or low production volumes)
- You have parts that are not built in very high volumes, but a family of similar parts exist that, when considered all together, make up a significant volume
- You build a product made up of several welded subassemblies and would like to weld one complete set of subassemblies at a time for just-in-time delivery to the assembly area

- You require higher quality than your welders are able to provide, or you cannot find and hire welders with the appropriate skill to weld at the level you require
- Safety of your welders is becoming a more critical issue
- There is discomfort or danger from the heat caused by preheating parts and by maintaining interpass temperatures

The list goes on, but if you have been reading this book thoughtfully, you could probably list ten more reasons why one might want, or need, to automate.

Step 2: Initial Vendor Inquiries

If you are confident enough to know that your application requires a robot, for example, then it will be easier to line up vendors, although a very large number of companies now call themselves robot suppliers. If you have no idea how your parts should be welded, then you may have to start with many more initial inquiries in order to cover the spectrum of automated welding types. Try to identify vendors who can supply robots as well as other types of automation. If a salesman only has robots in his pocket, then you can bet the farm that the only thing he can quote is robots, when you may really need a hardtooled system at one-third the cost.

What should you look for in a vendor? First, there should be those on staff who really know welding. As strange as it may seem, some welding machinery vendors do not have much knowledge of welding processes, consumables, or metallurgy. This is a very important and often overlooked criteria and one to which you should pay close attention.

Second, identify vendors who have been around to see what works and what does not, who have a good range of experience with the various types of automation, and who have first-hand knowledge of production and manufacturing. They should visit your plant, preferably on a regular basis. As much as possible, a vendor should strive to understand how to best fill your needs and help you become more efficient, not just to sell you a machine, and the only way to do that effectively is to be where the action is.

A good welding automation supplier will ask lots of questions. A poor one will immediately assume that he can solve any of the problems he sees. A good supplier takes time to understand where you are currently, *before* giving you a quote. A poor one will jump right in with a proposal. If you feel that the company you are dealing with just does not seem to understand, because they are not interested in spending the time, then be wary of their true agenda.

A good supplier will have a good reputation in the marketplace. They will give you a user's list of customers that you can call. Here, by the way, is an effective and revealing test: if an automation company boasts about the 1000 systems they have installed and the user's list that they give you has only 10 names on it, something might be fishy. Ask them for a comprehensive user's list. I don't think many companies will be willing to hand that information out for fear of bad press. Some will, however, and the basis of such vulnerability is the trust they have established with their customers.

If you do get your hands on such a list, you must still take the *negative* reports with a grain of salt. No one buys a car and expects it to run flawlessly throughout its whole life. You should not expect any more from a welding machine. If you hear bad reports, first of all be sure you understand what the problem really is, then find out how the vendor handled it. Having equipment problems is not against the rules. How they are addressed, however, speaks volumes about the supplier.

Be sure to ask the vendors and the users specifically about service, spare parts availability, and applications support, should these services be necessary. Of course, the vendors will be all roses when discussing their superior service capabilities, so find out from them how many service people are available, where they are located, and the like. Then find out from the users how the vendor reacts under real-life circumstances.

Depending on your product line, it may be a good idea to have all vendors sign confidentiality agreements. After all, you are trying to get the edge over your competitors, so you do not want to broadcast your decision to automate. More importantly, in the process of automating, you and your chosen vendor may develop new technologies or a new way of accomplishing some task, and you certainly do not want this information handed to your competitor on a platter.

At the same time, I recommend that you try not to be a pain in the neck about showing your system. If the vendor wants to publish an article about, or photograph, your equipment, encourage it. If confidentiality is a true concern, then it is usually possible to describe welding systems in such vague terms that it becomes unclear what the real purpose of the machine is. It is important to support efforts by vendors to spread the gospel about automating. This is needed if we are to keep our country's manufacturing edge in the world market.

Step 3: What to Tell the Vendors

Time frame

Identify the time frame you need to work in. If you are budgeting money for next year, then do not make the vendors jump through hoops. A ballpark or "budgetary" quote will give you all the information you need for now, and they are usually quite close to firm prices. Be aware that fixturing, as well as other options and services, can add significantly to the cost of a cell, and typically is not included in budgetary quotes. On the other hand, if you just landed a contract and need the system in 4 weeks, the supplier needs to know that as well.

Detailed drawings

Provide detailed weldment drawings and, if possible, component drawings. Some suppliers can take one look at a drawing and tell you that they are not able, or do not care, to propose a solution. This can save you a lot of time, so it should be among the first information to be shared. By the way, if you can put all of this information described in this section together in one organized package, it will be well worth your time, and you will make a lot of potential suppliers happy. Make sure you give identical packages to all who are quoting.

Visits to the facility

Encourage vendors to visit your facility. It is important for vendors to see and experience your production processes in order for them to understand the type of equipment that will best serve your needs. Product flow, production rates, the physical environment in which the machine would be working, and the philosophy of the operators are just a few of the things a vendor can learn firsthand in your shop, and all of these are important for them to know. While they are visiting, show them everything. The more they know about your company, the closer they will come to providing the ideal solution for your automation needs.

Budget constraints

I do not like to leave much room for budget constraints, but obviously some companies have a limited amount of resources with which to work. I hope you will purchase the right equipment for the job, not necessarily the cheapest. In any case, tell the vendors if you have an absolute maximum dollar amount that you cannot exceed. There may be more than one solution for your particular application, and the solutions can vary in price significantly. Sometimes it is possible to sacrifice 10% of the production capacity of a system for a 50% savings in the cost. If this is your goal, make it clear at the beginning.

Shop rates

Those preparing quotes should know your shop rates, broken down into labor and benefits, and overhead, if possible. Those who know what they are doing will use this information to provide a system with the best value for your particular production situation. It would also be a good idea to include your current costs for doing the same work manually. Again, if the vendor is truly concerned about providing you with the best solution, he can use this information to specify a system that is just right for you.

Some customers are adamant about not disclosing current cycle times, thinking that the vendor may use that information to skew his automatic cycle time calculations, making the system proposal more attractive. Do not worry about such things. All you have to do is make sure you understand how that engineer arrived at his proposed cycle times, then trust your own judgment about his calculations. You will be able to tell if he manufactured fictitious numbers and if his estimates are close to the real thing. The worksheets in chapter 4 can also be used to double-check cycle times provided to you. Besides, before you accept shipment of your system, you will witness a runoff, during which you can verify cycle times for yourself.

Current production method

Describe the current production method, including weld process used, weld procedures governing the manufacturing of the parts, the type of gas and wire required, and any other pertinent information you can muster. Remember, you are the experts on your parts. You may ask vendors for their opinions about welding wires, for example, but, in the end, it is your responsibility to choose the correct consumables.

Information regarding product flow can also impact the design of the system, so include a layout of its future location, indicate flow direction, and the amount of floor space available. The vendor should then, in return, provide a layout of the proposed system that fits well into your designated area and increases the efficiency of material transport within your facility.

Production rate

The required production rate is very important information. This can totally change the design and function of the system, and is very important for the creator of the proposal to have. Everything hinges upon this factor, including the type of automation needed, the attractiveness of the investment, and the amount of money you will need to invest. If production rates are very high, then a multiple number of machines may be needed, and the cost of each individual machine will greatly affect the total bill.

Team member list

Give vendors a list of all team members and their respective plant functions. If the vendor has a particular question about the welding sequence, encourage him to call the welder who knows such answers directly. If the questions pertain to power or utility availability, perhaps the plant engineer is the correct contact. Remain accessible, because the vendors may, indeed they should, have lots of questions in the early stages of the process.

Optional information

Tell vendors which other companies are quoting the job. This may allow you to gain insights into the relative strengths and weaknesses of each company. However, do not let it turn into a grudge match. A confident competitor will praise and respect his competition, not belittle them.

Step 4: What Not to Tell Vendors

Other price quotes

Not much is gained by telling vendors the prices that other companies are quoting. You may save a few dollars up front if they decide to cut and slash prices, but no one wins from such a situation in the long run. Welding is a competitive market, and by bargaining for a welding system, you are weakening the vendor's position in a field that we as a manufacturing nation cannot afford to lose; you are forcing them to cut corners, which may result in a less effective system for you, or vendors may be less able to support the project properly, since their profit margin was slashed too low. You are going to pay for your automated welding system in a year or two anyway, so an extra $10,000 in the up-front cost will make absolutely no difference in the long run.

Unique solutions from a vendor

If a vendor develops a unique solution to your problem that other vendors did not think of, or that is unique to that supplier's abilities, it is not fair to that vendor to ask others for quotes on the same concept. Chances are, since the idea was unique, and it addresses your application in a special way, the supplier has done it before, with success. At the same time, other suppliers probably have not done it that way before, and, if they tried, you might end up as a guinea pig. Unless the supplier is a proven custom equipment builder, it is usually safer to stick with proven solutions and technologies.

On the other hand, if the solutions quoted are generic in nature and something that any automation vendor should be able to do, then by all means have them all quote the same thing. This is the only way you can compare apples to apples. There are a lot of features that are unique, and a lot of good options available that can make life much easier. It is easiest to compare proposals if the equipment proposed is somewhat the same.

Interlude #1: What Vendors Will Tell You

Let us take a break from the routine now and delve into the complex and nerve-wracking world of automated welding marketing. These are some of the more common statements you will hear as you begin developing your projects. They are discussed here simply so that you will expect them and will know how to respond.

"Our robot is the easiest on the market to program."

All robots are easy to program. The amount of time it takes to write a program is rarely significant. What you can accomplish *during* programming, through the power and ability of the controller, is very significant. You do not purchase a robot to program it. Its purpose is to weld parts. The goal is to get the programming over with, so the investment will begin making money. By simply striving to minimize the programming time, you will be minimizing the importance of programming gimmicks, which may be quite impractical.

"You really need these ac motors and absolute encoders."

The arguments for ac motors and absolute encoders are often overstated: ac motors may last a bit longer than dc motors, but there are dc drive robots out there that have been working flawlessly for many years. Unless you have a very sound reason to choose one or the other, do not let this factor get in the way.

Absolute encoders can be nice, since they eliminate the need to synchronize, but how many times does a user really need to synchronize a robot? Only upon a complete loss of power, which is not that often. Although power loss from any number of reasons—power spikes, lightning strikes, someone tripping over the cord—can mess up a program, the fact remains that it normally takes only a couple of minutes to recover. Do not base your decision solely on this factor, since the end result will not significantly alter the productivity of the cell. Productivity and value are the key ingredients to look for.

"Our robot can control up to 12 axes simultaneously."

Most robots available today can control up to 12 axes. However, there are a few that cannot work with programmable servo axes, so be careful. Indexing positioners can be less expensive, and often work quite nicely in many situations. However, in many instances, programming them is much more cumbersome, and they are not as flexible as programmable servo positioners. If you think the extra flexibility of programmable positioners is what you need, either now or in the future, then be sure the robot controller has that capability.

"We are a true one-source supplier."

It is actually quite important to utilize a one-source supplier, and I very strongly recommend doing business this way. The benefits are numerous and dramatic. However, a lot of suppliers who are not true single-sourcers say that they are, so you need to check them out very carefully. There are very few who can truly single-source all of their equipment, including the welder, positioners, tracks, robot, welding torch, and the rest. There are distinct advantages in dealing with those who do—interface problems are usually nonexistent; there will be no finger-pointing if a particular system component fails; and there is one source for spare parts, service expertise, and applications help. Do not just accept this line if you hear it. Investigate for yourself, and see exactly what your vendors mean by "single-source."

"You need a robot for this job."

If a robot supplier calls on you, that person will want to sell you a robot. If a salesman peddles hard automation, he will try to put that in your plant. In either case, if you buy the wrong machine, you will regret not taking the time to find the correct solution. You have already learned of the various types of automation and where they fit. Use this information to weed out those who are simply trying to sell a machine from those who are truly interested in helping you achieve your goals.

"We've done this before."

A potential vendor may tell you of similar systems they have done. This information is like gold. First, it shows that your application can indeed be done automatically, so the risk is reduced. Second, he may actually offer to take you to the location to see the machine in action. Go! Third, it may give you some insight into how your competition is doing it. In any case, ask to see the system in action, ask for a contact at the company that you can talk to, and ask for pictures and other information about the system. If the vendor comes through on all these requests, and you follow up with the customer, then you will know exactly where that supplier stands in his ability to meet your specific needs.

Interlude #2: What Vendors Will Not Tell You

I may be going out on a limb here, but it is very unlikely that you will hear the following statements from your automation supplier:

"Would you like to see our list of customers who are unhappy with the systems we have provided them?"

Suppliers don't like to talk about their mistakes. I don't know if there is much value in forcing them to tell you this information, since there is no such thing as a perfect company. Everyone has skeletons in the closet, but what was done with those old bones, and the status of the company's overall track record, are what is important.

"This job is too much for us. We don't want to pursue it."

In reality, I do know of a couple of companies who have used this line, and I have great respect for them. It is not common in today's sale-hungry world. It is up to you to weed out the pie-in-the-sky ideas and identify the vendors who can really do the job. They will not generally like to talk about their limitations. In fact, some vendors seem to prefer taking on jobs that they cannot handle, then losing money on the project, missing shipment dates, and generally wasting their customers' time and money. Try to visit existing customers' plants to see the vendors' ideas in action. There may be times when your solution requires delving into some type of development work, so there will be no users, since the equipment will be unique. However, you can find out a lot about a company by visiting their customers and asking the right questions.

"Have some of our competitors quote this same solution. See if they agree that it will work for you."

It is up to you to seek out other quotations, and it is up to you to decipher whether a vendor is giving you a realistic solution to your particular application. If you have any doubts about a particular solution, ask other experts about it.

Step 5: Properly Define the Project

Now, back from our digression. It is time to discuss the importance of up-front communication regarding proper definition of the project. All parties involved should know exactly what the proposed system will do, which welds it will complete, how it will operate, and how it will be built.

I have personally seen this happen; an automation company built what they thought the customer needed, the customer had a completely different idea of what he was buying, the time for the runoff rolled around, and the sparks flew. All of this could have been avoided with a little homework. Your automation team needs to understand exactly what the proposed system will be expected to do, exactly which welds it is expected to complete, how fast it can complete the cycle, and precisely how it will function on a day-to-day basis. During project development, the team should be in close contact with the supplier, so they understand how the machine is being built. If a change needs to be made, the time to do it is at the supplier's factory. Sometimes a user may see something he would like to change while the machine is being built. This is acceptable, but expect to pay above and beyond the quoted price if it requires a major design change that was not described in the purchase order. Remember, if you just give them your purchase order, and then you forget about the project until the runoff date, you could be heading for trouble.

Following is a list of things you may want to keep tabs on. Some may be quite obvious. Others may seem obvious, but are still commonly overlooked. Still others may be things you just hadn't thought of before. They are all good questions to ask while you are defining your project, and all can help alleviate future headaches.

1. What type of basic machine is involved? If it is a robot, is it a five- or a six-axis model? What brand is it? Some integrators represent many robot brands. If it is a dedicated welding machine, who builds it? Sometimes you will deal directly with the builder, sometimes with a representative, who may or may not know a thing about welding and automation.

2. How will the system function? Are operator's panels included that allow cycle start, safety reset, E-stop, cycle stop, and any other functions you may desire? Is loading and unloading internal to the welding process, if desired? Are changeovers quick and easy? Ask for a demonstration of these tasks.

3. Is memory adequate? Any programmable machine will require a certain amount of memory for all your programs. Make sure there is more there than you think you will need. A robotic welding program, for example, requires a minimum of three points stored per weld; the weld start point, the weld end point, and the approach point. If the weld is not a straight line, but must turn a corner or form an arc, then more points are required. It is easy to

see that a program with 60 welds will require at least several hundred instructions. If you want to store ten such programs in memory, then be sure that not only available RAM memory is adequate, but also the memory on the floppy drive or other mass storage device.

4. Do you need more than two stations? Sometimes, the addition of additional workstations can cost very little, compared to the whole system price. This small investment now may eliminate a lot of downtime for changeovers in the future.

5. Is bulk welding wire a possibility? If so, will the wire travel properly through the additional length of conduit? When you start pushing welding wire through longer liners, feed problems can cause severe defects in the welds. This type of feed problem is quite common, especially with larger diameter, stiffer wires.

6. Are ways and bearings protected from weld spatter and drops of molten metal? Are wires and cables also protected in conduit or under covers? This especially applies to welding fixtures, since clamps, bearing surfaces, and sliding mechanisms are particularly close to the welding, and quite vulnerable to pitting, scarring, and melting.

7. Is safety properly addressed? When you look at a vendor's concept, look for ways in which an operator may be injured or placed in danger. Rotating and sliding mechanisms should be well covered. Be sure that welding fumes are properly addressed. Arc flash should be well guarded wherever possible. Other unsuspecting employees should not be able inadvertently to walk into the welding cell. Are E-stop buttons located at convenient places throughout the cell?

8. Will maintenance be easy to perform? Mechanisms, though covered for safety, should be easily accessible. It should be easy to change consumables. Wiring should be well labeled, and hydraulic and air lines should be neat and well thought out. Are components common and readily available? You may want to specify a certain brand of PLC, valves, or other common parts that your maintenance shop already stocks.

9. What type and brand of welding power source do you want? Many are available, and some offer technology that is unique and beneficial. Many types and sizes of welding torches are available as well. Larger torches can limit your access to certain welds, but smaller torches may limit the amperage or duty cycle you obtain from the machine.

10. Are peripherals such as servo tracks or positioners necessary? If the welding torch cannot reach certain welds, it may be necessary to bring those welds to the torch by manipulating the weldment. How exactly is this external equipment programmed? Is it a laborious process of I/O communication, or is programming done through the machine controller?

11. Is fixturing adequate? Although easily accessible to a human being, certain welds may be quite difficult to reach with a less flexible machine. Be sure that clamps are not in the way, and, as silly as it may sound, be sure that the part can be removed after welding. Other than the obvious, a common problem is that weld distortion puts stress on the parts such that pins cannot be pulled out, or clamps cannot be removed. Please remember, though, that even the best fixture designers and builders must at times take the trusty disc grinder to a particular fixture detail in order to make room for the welding torch.

12. Are pneumatic lines resistant to the heat and the abusive atmosphere of welding? Are controllers and other electronic components protected from sparks and spatter? Are they industrially sealed from dirt and contamination? If you are in a warm, humid climate, does the controller have an air conditioner to alleviate overheating problems? Are proximity switches and other delicate sensors properly protected?

13. If TIG welding or plasma cutting are used, are critical controls hardened against the high frequency signals created by such processes? Some equipment requires specific optional accessories in order to be shielded from the harmful effects of high frequency.

14. Precisely which welds is the machine designed to reach? Accu-

rate definition of the runoff requirements will be well worth the effort it takes to discuss and write them down. How much additional assistance is included? Will the supplier be there until the system is production ready or only long enough to install it? Either situation may be correct for your application; just be aware that it must be defined in the beginning.

You may be able to add to this list. It is intended to get your mind thinking correctly about the level of involvement you and your team should have in the early stages of the project. Just as weeks of preparation are needed to compete in a drag race that lasts for five seconds, you can insure your future success by doing your homework now. Have a firm handle on your solution before any equipment is even built.

Programming a Robot

As far as the actual writing of a welding program goes, most robots are programmed in a similar manner. A teach pendant is used, which is a small control box attached to the robot controller by a long cable. The operator actually instructs the robot controller that he is ready to write a program, and then takes the teach pendant with him to the welding fixture/robot area. The controller may be 20 to 30 feet or farther from the robot itself, so the length of the teach pendant cable is important.

A robotic welding program is basically written this way:

1. The robot is "taught" to the desired point. The most common method of moving the robot is with a row of 12 push buttons (for a six-axis robot)—a "plus" and "minus" key for each of the six axes. By pushing these buttons, the operator teaches the robot into the exact location desired. The first point programmed is usually a "home" position, which will be a common safe point at which all programs will begin and end. When the robot arm is in the proper position, that position is stored by pushing the appropriate button. This will store the resolver or encoder counts for each of the six robot axes. If the robot is mounted to a programmable

servo track, the track position is also stored. If the welding cell includes programmable servo positioners, these positions, too, are stored. With a single push button, up to 12 axes of motion (in most robotic systems) can be stored simultaneously.

2. The robot is then taught to the beginning of the first weld. If fixture components or the weldment itself is in the way, several points may need to be stored to provide a path around the obstacle. A robot will move directly from point to point, not knowing whether anything is obstructing its path.

   Most robots have more than one means of approaching a point. First, they can be instructed to simply take the fastest route to the point. This means that the robot does not care whether it moves in a straight line, it just takes the fastest path, which usually means the path that requires movement of the fewest number of axes. Second, a robot can be instructed to follow a Cartesian coordinate system, so it can interpolate a straight line between two taught points. The Cartesian coordinate system is always used while welding, to guarantee that the robot will accurately follow the desired path. A robot can also interpolate a circular arc through three taught points and can mix straight lines with arcs.

3. At the weld start point, an instruction is given to the robot to turn on the welding arc. This actually consists of a sequence of events, including:

- Gas preflow, to displace contaminating air around the weld zone with shielding gas
- Arc start sequence, which guarantees a consistent arc start
- Current sensor monitoring, to ensure that the arc has started before the robot begins traveling
- The activation of the main weld parameters, which are the welding voltage, wire feed speed, and amperage to be used during the main part of the weld

   The correct torch angle must be programmed by manipulating the robot wrist. This is done by rotating the torch about its tool center point (Figure 5-1). The proper torch angle must be defined at the beginning of the weld as well as at the end of the weld. Linear

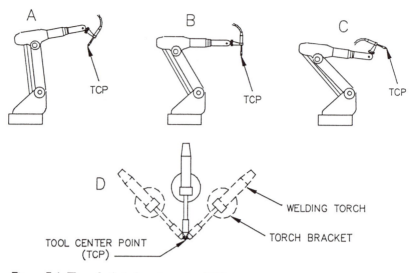

Figure 5-1. The robot's tool center point (TCP).

interpolation should maintain the torch angle during the weld itself.

4. If intermediate points are needed during the weld (if the weld travels through an arc or turns a corner), then these are programmed as well. Again, the torch angle must be updated by the programmer at these intermediate points, or the robot may not maintain the correct angle.

If positioners are part of the cell configuration, then they must be programmed, too. Programmable servo positioners are quite simple to program. The operator simply teaches the positioner into the desired location, then teaches the robot into its desired location, then presses the "store" button. Since storing a robot position automatically stores the positions of all servo axes connected to the system, it is a simple and straightforward procedure. On the other hand, if the system utilizes indexing positioners, the programming can become more involved, since the operator must program individual commands to cause the axis to index, then additional commands are needed that check inputs to verify that the axis has indeed rotated as desired.

5. At the end of the weld, the robot position is stored, and the instruction is given to turn off the welding arc. This may include a ramp down of welding heat, and/or a crater fill function. The crater fill is designed to extinguish the arc and pause for a preprogrammed length of time to allow the weld puddle to solidify, then to restart the arc long enough to fill the crater completely. These cooling times and crater fill times are programmable by the operator and depend on the type and size of welds being programmed.

6. The robot is then programmed to travel to the next weld, and it starts all over again. A complete robotic welding program consists simply of many individual points strung together in a program that, when run, executes these instructions in a smooth, continuous fashion. When the welding is complete, the robot should return to its home position to wait for further instructions from the operator.

Summary

Like most welding manufacturers, you may have a little homework to do before a welding automation installation can be successful. Here is how to minimize the risk and maximize profits.

A. Vision

1. You must first be convinced deep in your heart that you are truly committed to welding automation.

B. Operator involvement

1. The welders are the experts. Tap their knowledge, use their expertise, give them power to make decisions.

C. Management commitment

1. True commitment differs from interest or curiosity.

D. Organizing the team

1. Who should be involved?

a. Management

b. Design engineering

c. Production engineering

      d. Quality

      e. Welding operator(s)

E. Step 1: Identify the project

  1. Identify parts that are regularly produced.

  2. High part volumes may indicate a need for hard automation, small part volumes may require the flexibility of robots.

  3. High weld volume per part may require robots.

  4. Families of subassemblies may be good candidates.

  5. Higher quality standards or lack of skilled welders may require automation.

  6. None of the above rules may apply to your situation. Welding automation can be complex and unique.

F. Step 2: Initial vendor inquiries

  1. Contact several vendors; as many as six may be helpful.

  2. Contact some who specialize in robotics, some who specialize in hard automation, and some who are experienced with both.

  3. A good vendor should know welding.

  4. A good vendor will do more listening than talking.

  5. A good vendor will have a good reputation.

  6. User's lists can be like gold.

  7. You may consider confidentiality agreements.

G. Step 3: What to tell the vendors

  1. Identify your time frame.

  2. Provide detailed weldment and component drawings.

  3. Tell vendors your budget constraints.

  4. Tell vendors your shop rates, overhead, and current weld process information.

  5. Describe your current production methods, including nonwelding processes.

  6. Tell vendors your desired production rates.

  7. Give vendors a list of all team members, and encourage them to call individuals on the team directly.

H. Step 4: What not to tell vendors

  1. Other vendors' quoted prices

  2. Other vendors' unique ideas

I. Interlude #1: What vendors will tell you
1. "Our robots are the easiest to program"—(They are all easy to program)
2. "You really need ac motors and absolute encoders"—(Not necessarily)
3. "Our robot controls up to 12 axes"—(Most do nowadays, however, a few still do not)
4. "We are a true one-source supplier"—(Really? You should check them out thoroughly)
5. "You need a robot for this job"—(Perhaps not. A robot may be more expensive and slower than the correct solution)
6. "We've done this before"—(Make sure they can prove it)

J. Interlude #2: What vendors will not tell you
1. "Here is our list of unhappy customers. Call them."
2. "We do not want to quote this job. It is too much for us."
3. "Just to be sure we are not pulling your leg, have some of our competitors quote this same solution."

K. Step 5: Properly define the project
1. Proper definition of the project is vitally important.
2. Ask all of the questions shown, and more.

L. Programming a robot
1. Be sure to understand how a robot is programmed. Then you will not fall prey to marketing gimmicks, and will understand exactly what your startup will entail.

6. Narrowing the Field

At this point, you should have a good handle on what projects to pursue, roughly how much it will cost, and what your savings will be. However, you have no idea who these suppliers are that have quoted these various, and possibly quite diverse, solutions to your welding automation needs. You need to find somebody to trust, someone who has the experience and engineering expertise to provide a system that will really work, and someone who will support your efforts, not just drop off the machine and bid "good luck!" on their way out the door.

Evaluate the Vendors

Separating the wheat from the chaff can be rather time consuming, but the automated welding vendor you choose will either assure success or compromise, depending on his character and abilities. Answers to a few pointed questions will get you well on the way to a successful partnership.

How long has the vendor been in business?

The new kid on the block may have some great ideas. In fact, they may be the supplier best suited for you. However, it is hard to argue with success. If someone shows up who has been around for 10 years and has a history of successful installations and a reputation in the industry for installing systems that work, then you can be reasonably sure they will do your job well. This question is important in considering any machine purchase but is particularly important in automated welding, especially when discussing robotics, since it is a relatively young industry. There are not a lot of suppliers out there who will do a great job every time. There are a few, and it is your job to identify them.

How many systems has the vendor installed?

Again, younger companies can sometimes provide much more innovative solutions, but, without the expertise that past experience provides, there could be some snags. However, do not compare sheer numbers of installations, thinking that the one with the most experience wins. The purpose here is simply to verify whether it is a fly-by-night organization or whether they have been around long enough to install some systems successfully. A better question might be, "How many of the systems that this vendor has installed were successful, and how many are still running today, as intended?" Many early robotic welding systems were mothballed, just as often for failures by the end user as for the suppliers' shortcomings.

Is the company financially stable?

You may be tempted to jump at a bargain-basement price, but be aware that, in all businesses, that can be a sign of instability and desperation. "You get what you pay for" is as true in the automated welding field as in any other. If that bargain seems too good to be true, it probably is. The machine may well be capable, but you will feel the effects in after-sale support, spare parts availability, applications support, and in myriads of other ways.

Do personnel or other physical resources seem adequate?

When considering robotic welding, be aware that fewer and fewer robot manufacturers are acting as their own integrators. Some do indeed take full responsibility for integrating their own complete welding cells and use their own robots, welders, positioners, welding torches, and everything else in the cell. Many others sell through integrators, though not necessarily exclusively. When acting as integrators, it is important that the company that is expected to build, install, and service your welding system is well prepared for the job, with adequate facilities, skilled workers, and a good support system.

The integrators buy the cell components from the robot company, then assemble them on their own floor, adding tooling, if necessary, and providing any additional equipment or services to complete the package. These integrators are usually quite capable of providing robotic welding systems that work. The question that remains is what type of work do they do best? Some are particularly well suited for welding smaller weldments that do not require much extra positioning during welding. Others are best at medium-sized weldments, but that still do not require very large memory capacities or sophisticated controllers. Still others specialize in attacking large, heavy weldments, or projects that are technically more challenging than the average. If you mismatch your project with the wrong supplier, it can lead to some headaches. By thoroughly following up on the user's list they provide, you will be able to discern what type of projects each vendor is best suited for.

Will the company's service and support be adequate?

This item is altogether as important as it sounds. The biggest complaints heard about some automation suppliers deal with spare parts and service availability. If you rely on that machine to serve up a predictable batch of finished parts every day, you cannot afford to have the machine down for a week waiting for spare parts to be shipped from across an ocean. Spares should be available, and service should be prompt and effective. How can you find out these things? Talk to end users. Most automation companies will give you some references to contact. Call and ask them brutal questions about service and support. Your success is on the line.

User's lists

User's lists are very important. You can tell a lot about an automation supplier by the user's list they provide for you. If a company lists every one of their customers and encourages you to call any of them, then do it. If you have time, call every one of them. Ask about the startup; about the support they received; if they would do it all over again the same way. Ask whether the installation was successful, if the system is making money for them, and if they will buy their next system from the same vendor. Be careful, however; as I mentioned before, if a company boasts about their hundreds or thousands of installations, but the user's list they provide has only 10 names on it, then you might want to look into it a little deeper.

Evaluate the Proposals

Evaluating the proposals can also be quite time consuming. That is why I recommend generally no more than four or five initial inquiries, unless you have some time on your hands and can afford to consider more than that. The proposals you receive may be significantly different from each other, especially if you have a good cross section of companies, for example, some that deal in only hard automation, some that handle only robots, and some that provide both. Even if you are convinced that your application is a robotic application, and all the proposals are for robot cells, you could still end up with quite a variety of system configurations. How do you know which is the best for you?

Some systems will lend themselves more toward flexibility, perhaps at the expense of productivity. Other concepts might be geared for full-speed-ahead high production, with not much hope of using the equipment for other purposes in the future. Various proposals may differ drastically in cost.

You need to know what is important to you. If a simple, inexpensive machine provides a payback of 6 months, it would be hard to argue for a robot, unless you know that it can be effectively retooled after the current contract is over. I hope you are familiar enough by now with the pros and cons of the various types of equipment, and know what you want. That will make the job at hand much easier.

It is possible that all of the chosen vendors can do the job effectively, even though the solutions may not be similar in design or function. Hopefully, the following information will be included in the proposals on your desk and will help to answer these questions. If not, make sure you know these answers. After gathering this information, your goal will be to narrow the field down to two or three candidates.

Exactly how will the system function?

1. If the machine is welding at Station One, and the operator has finished reloading Station Two, he should be able to push the "Ready" button and walk away. The signal should latch on until the welding at Station One is completed, and the welding should then immediately commence at Station Two.

2. Emergency stop buttons should be located at convenient points throughout the cell. If the E-stop is pushed, all system functions should immediately stop. Most importantly, it should be a quick and easy task to recover from E-stop and resume welding.

3. Multiple stations should almost always be quoted in a robot cell, except in the case of very large parts with long cycle times or other special applications. Multiple stations are also feasible in many hard-tooled applications, by using indexing tables, conveyors, or other means of indexing parts. Multiple stations ensure that loading and unloading take place internally to the welding cycle, which is one of the main sources of justification (see chapter 4).

4. All welding automation should include comprehensive controls for adjustment of all pertinent weld parameters. The machine was purchased for welding, so demand that it speaks to your operators in welding terms and provides the necessary adjustability. For MIG welding with hard automation, the absolute minimum includes controls for voltage, wire feed speed, travel speed, and burnback (to prevent the wire from freezing into the puddle). Many controls also include pre- and postpurge gas controls, slow wire feed run-in, crater fill, initial control for hot starts, and others. These can all be useful at times, depending on the level of sophistication that your welding application requires.

   TIG welding with hard automation can require much more sophisticated controls. Controls are available, for very high-precision TIG and plasma welding jobs, that control all of the above parameters, plus many others such as: cold wire feed speeds and wire feed pulsing; current pulsing; voltage regulation for precise standoff heights; gas flow rates; upslope; downslope; and pulse parameters, such as pulse frequency and pulse width. The suppliers of such equipment are normally very well versed in the use of and need for such controls, so discuss these issues carefully with them.

For robotic welding, be sure that the weld parameters are controlled through the robot software. This gives the user the flexibility to call various weld schedules automatically in the robot program, change weld schedules "on the fly" (in real time), and quickly and easily edit weld parameters and weld speeds.

5. On hard automation, torches should be adjustable, especially with flexible hard automation, where setups will be frequent. The machine should also include some method of zeroing the welding torch to a known location, possibly a separate datum point for each part configuration you expect to weld.

Robotic welding systems should not have adjustable torches but should be equipped with a means of resetting the TCP (Tool Center Point, the point in space that the robot thinks it is moving around, usually coinciding with the tip of the welding wire at the proper stickout). This is especially important after a crash (yes, you will crash the robot and not just once) or after the torch has been changed or moved for any reason.

Herein lies one of the most confounding problems in automated welding. It does not take much of a crash to move or bend the welding torch slightly. When this happens, I have seen people swear that they just got a bad batch of parts in the cell, and that is why the welds are mispositioning. In reality, the TCP is off because of the crash, and they need only readjust it to get back on track.

6. Is it easy to change consumables? Will the operator be balancing on one leg on a ladder and holding onto the weld curtain with his teeth, while trying to change the 60 pound spool of welding wire in mid-air? I have seen systems like this. The machine should be user friendly when it comes to changing consumables, contact

tips, torches, torch liners, and such. A bulk wire dereeler can be a very nice addition. Bulk dispensers can pay for themselves in just a few months.

7. Is the system safe to operate and to work around? Guarding from moving parts during the weld cycle is a must. Dual palm buttons for cycle start are a good idea. If possible, arc shields, which automatically close as the weld cycle begins, should be installed, and the rest of the machine should be shielded so that stray welding arc is not dispersed all over your shop. Realize that, on many applications, it is simply not possible to shield the entire cell completely, but you should demand that such safeguards be installed where possible.

If smoke is a problem, most automation suppliers can fit the welding system with some type of fume removal or filtering equipment. This is becoming more crucial as standards become more strict. This is especially important when welding with flux cored wire, when welding on dirty or oily parts, or any other process that produces a lot of fumes. The filtering system will be a small part of the total cost of the cell, so invest in clean air. Everyone will benefit from it.

In short, you should be aware of every function of the machine, since you will be living with it for the next several years. If there is anything you do not like, change it before they ship it to you! Obviously, if the changes are substantial, the vendor has the right to charge you for them.

Ask about similar installations

If the supplier has built a bunch of these machines, then they probably know how to do it by now. If it is a one-of-a-kind effort, then be careful. Obviously, there are those companies that specialize in custom welding systems, and many do excellent work. I have stressed this before, but it is important that you contact other customers. Especially concentrate on systems that would be similar in design and function to yours, if that is possible. Not only will this assure you that the supplier can actually do the job, but also, perhaps more importantly, you will see for yourself—and you will be showing your automation team members—that it actually is possible to do it that way, and that there are people out there doing it successfully. This can add a lot of momentum to the project.

The (almost) final cut

After considering the other factors that we have already discussed in this book (future flexibility, productivity, company track records, etc.), and assuming that the delivery schedules will work for you and the terms are under control, you will be ready to make the final cut to two or three (preferably two) finalists. Use your best instincts, armed with the information you have been gathering, to narrow the field, but be sure you know exactly why you are eliminating some and holding on to others.

A quick word about delivery schedules. Deliveries should not be a huge priority unless, of course, you have just landed a red-hot contract. If you find the right solution to your welding problems, but the lead time is long, do not accept a substitute for your Romeo or Juliet. "Place the order." You have survived for years up to this point; chances are, a few more months will not hurt too badly. Besides, you will need some time to get your house in order, too.

Sell the Project Internally

This is a good place to discuss the importance of perception. Automated welding projects seem to raise a lot of eyebrows, and much attention is paid to those spearheading the project. By much attention, I mean this kind: "I hope he fails so I can say 'I told you so.'" Or "We tried robots here before, and it didn't work. Why does she think it will work *this* time?" I hope your company is not like this, but I must be frank. Welding automation is not easily embraced in this country. The statistics prove it. I want you and your team to be the welding success story in your plant. I want you to be the heroes, and you can be. I do not want to spend a lot of time here, but we need to discuss just a few ideas that might help you get some momentum rolling.

Do not oversell

If you waltz into the boss's office and announce that you can save him $100,000 in 1 year with a welding robot, you better be able to prove it. If you tell him you can cut manual welding labor by 50%, you better be able to show savings of 60%. If you have to back up in your tracks because of a miscalculation, the project could be doomed before it even starts. Usually, just stating some numbers that you know are very conservative will be enough to impress the right people.

Speak in generalities

The project could change 15 times before a good solution is found. To avoid using a lot of white-out, talk in general terms, like "A welding machine could weld our yearly volume of this part in less than two shifts" or "This system would address the welding fume issues that OSHA is pointing out to us." You will have plenty of time later to fill in the specifics.

Emphasize the positive

I had the privilege of reading a letter that was written by an engineer whose job it was to determine whether welding robots would be a useful investment. After three pages of laborious explanations, he boiled his analysis down to some pros, some cons, and some neutrals. Pros included improved productivity, and improved quality and safety, and heavily outweighed the negative aspects of the purchase. A less than 2-year payback seemed to ensure that the prognosis would be in favor of a robotic welding system. The last paragraph, however, explained that he thought the money could be better invested elsewhere.

The thing that stood out in my mind about this letter was the item, listed as "negative," which stated that they would have to train their welders to become robot operators. I wondered to myself "Since when is training people and increasing their worth as employees a negative thing?" Other statements in the letter indicated that, beneath it all, this person was obviously scared to death to make a decision about automation. He minimized the gains to be had and emphasized the problems that he envisioned—some of them completely fictional—in order to avoid making a decision.

There are plenty of positive things to say about welding automation if you have the right application. Do not be timid. Do not oversell it, but be truthful and accurate in explaining the benefits that can be derived from automating.

Include overall benefits

This simply means that there are more benefits to be had from automated welding than first meet the eye. We discussed these thoroughly in chapters 1 and 4. Productivity and quality will increase—you have heard this several times already. You have also heard that other benefits are often derived from such automation, such as less spatter, which leads to better looking parts. You may reduce subsequent operations by automating, such as grinding or straightening. Or, by improving part fitup and eliminating gaps, you are building a better overall product, with stronger weld joints and tighter tolerances. This could allow you to decrease weld sizes, providing huge savings in welding consumables. Worker fatigue may decrease, new customers may be won, existing customers may send more work your way, and you may eliminate welding bottlenecks. Enumerate these factors—if they are realistic and easy to prove—in the initial descriptions of your project to give it more weight and to make people sit up and pay attention.

Include and credit the entire team

This can bring the team together more effectively, it will give the members a sense of pride to know that the big boss is looking to them for answers, and it will force accountability from the team. When all eyes are focused on you, it is uncanny how decisions are suddenly made with great care. If this welding automation swat team indeed functions as it should, lean and mean, you will be surprised at the incredible ideas and productivity that will flow from it.

Final Proposals

Now let us get down to the nitty-gritty. Here is where we dissect the final proposals and choose our vendor. Remember, these are only guidelines.

If possible, you should compare apples to apples. For example, the choice may be between two similar robotic solutions. If this is so, then you will have to delve a little deeper into the histories of the respective suppliers. Sometimes, you may end up with two wildly different solutions, either of which might do an equally effective job. In this case, you will be comparing apples to oranges. This is not always an easy situation to be in, but rest assured that there may be more than one right answer.

Welding demonstrations

Welding demonstrations may be a part of your decision-making process. Many potential customers require demos before even considering a welding system. This may or may not be appropriate behavior. If your job has been done elsewhere with great success, and the vendor can effectively convince you of it, then a demo is probably not necessary. Besides, you will demand a functional runoff before approving shipment anyway. If the system does not work as described, then it is not shipped until you approve. Make sure this provision is clearly defined in your purchase order.

However, if the proposed solution requires developmental technology or the vendor simply has not built such a machine before then asking for a demo makes good sense. Also, if your welding application is particularly challenging for whatever reason, then a demo may be the only way to identify those vendors that can do the job correctly. Be ready to pay if the demo requires extensive time or material from the vendor. At this point, there is no guarantee that you will purchase a system from any particular vendor, so they need to minimize their costs. This is just good business sense on their part.

Factory demonstrations typically include the building of some simple tooling, programming of the machine, and a runoff of sorts to prove that the machinery and/or the vendor can do the job. Do not expect perfect results. The demo equipment is typically lab equipment, so it may not be the optimum design for your application, and the fixturing will not be production tooling. The purpose here is to study the process, see if the vendors know what they are talking about, and achieve some comfort level that will allow you to pursue a relationship with that particular vendor further.

Factory visits

The next step will be factory and user visits. It is important to visit the factory in order to get a little information on who the vendors are and how they operate, and, obviously, to view a demonstration, if this is applicable. Take a good look at their production area. Do they have appropriate resources? Ask to see their spare parts facilities and service department. Do they keep major components in stock? Are their subsuppliers well stocked? How long will it take to get a serviceman to your facility? Are engineering resources adequate? If possible, choose a vendor that does all of their own engineering and development work. One-source responsibility is not just a buzzword. It can be the difference between life and death in a welding project.

More important than visiting the factory is visiting end users. After all, as long as the machine gets built effectively, it does not matter much where it was built. The test comes after installation on the customer's floor. If it works there, and the customers are happy, and if the customers have honored the vendor with repeat business, and if the users tell you that they would buy another system from the same vendor, then you are onto a winner. Talk to the people who were in charge of the project in its initial stages (your counterparts). Ask the questions that are on your mind. Be open and honest. After all, you are about to invest a very respectable sum of money. Get all the information you possibly can.

Talk to the operators who are running the equipment. Ask them how long it took them to learn their way around the cell and how easy or difficult the system is to use. See if they have any suggestions on what they would do differently if they had to do it again. Also, see if they would prefer to go back to the old way, sweating under a hood and leathers all day long.

Talk to the maintenance people, and see if the machine is easy to troubleshoot. Make sure that what the vendors told you about parts and service is accurate. Talk to the managers of the plant to get their opinions on the feasibility and/or profitability of their automated welding system and to see if they have had any regrets. Do not be afraid to spend some time, if they will allow it. Chances are, most end users will be proud to show off their success and will be ready and willing to spend the time that you need.

While at their factory, sit down at the conference table with your vendors and discuss their final proposals in detail. Make sure you understand every line and what takes place at every step along the way. It is not a bad idea to ask for a schedule or Gantt chart on the production and delivery of your system. Although I do not recommend playing hardball on deliveries, you should at least expect to receive the system when they say you will.

Be ready to contribute something to the project. Just like in a good marriage, 50/50 is not enough. You must each give 100% to ensure success. This analogy is appropriate now, since we are about ready to delve into a long-term relationship akin to marriage. The following chapter will discuss exactly what it takes, now that you have chosen the correct vendor, to make it all come together. Finally, with all of this ammunition at your disposal, take the leap, and choose a vendor.

Summary

This is probably the most important decision to make: with which vendor should you form a long-term relationship? Most suppliers are capable of building machines that will move the welding torch from place to place. Competition is stiff regarding functions and features. The important criterion is the company supporting your automation efforts.

A. Evaluate the vendors
   1. How long have they been doing this?
   2. How many systems have they installed? Sheer numbers are not as important as how many of those installations are still working today.
   3. Are they financially stable?
   4. Are there adequate resources?
   5. Is the company's service and support adequate? What is the opinion about them in the field? Are spare parts readily available?
   6. Get a user's list. Use it. Analyze it.

B. Evaluate the proposals
   1. Get a variety of proposals from a variety of companies. If your product lends itself to more than one type of automation, get quotes for those various concepts.
   2. Determine exactly how the machine will function. If such information is unclear, and cannot be articulated by the supplier, chances are they do not know what they are talking about. A few basics:
      a. Controls should be user friendly.
      b. Convenient E-stops are important.
      c. Internal load/unload is a high priority.
      d. Controls should be comprehensive, allowing adjustment and monitoring of all important parameters.
   3. Ease of maintenance and safety are important issues.
   4. Ask about similar installations.
   5. After considering everything in this chapter, as well as terms, delivery, and other conditions, make the final cut to two, perhaps three, finalists.

C. Sell the project internally
   1. Don't oversell. You probably will not need to.
   2. Speak in generalities, until the project is well defined.

3. Emphasize the positive, but be honest about the drawbacks (if there are any).

4. Include overall benefits, not just faster weld cycles.

5. Include and credit the entire team.

D. Final proposals

1. If possible, compare apples to apples.

2. If solutions vary considerably, remember, there may be more than one correct solution.

3. Visit the factory and visit end users. Ask lots of questions and spend as much time as possible.

4. Be sure you understand every line of the proposals, and be sure you know exactly what you will be getting for your money.

7. A Match Made in Heaven

Just like in a good marriage, a few lessons must be learned to assure success: listen a lot; give 100%; be honest, open, and vulnerable; develop a relationship built on trust. (To address undying love and devotion might be taking this analogy a bit too far.) In many ways, your relationship with your vendor should be the same. The name of the game is teamwork, and it is absolutely essential in order to prosper.

The Courtship: Excitement and Nervousness

You have chosen your supplier and have placed the order! If it is your first automated welding system, you may have a few jitters. Your boss is applying pressure for you to succeed, your peers are watching your antics, and your own pride wants you to succeed. The team has done their job well, yet it has only just begun. These next few weeks or months may be a time for you to relax a little, but many important activities must take place before you are ready to accept shipment of your equipment.

The engagement

It all starts when you send the order to your selected vendor. Some companies prefer to copy the vendor's quotation verbatim and use that as the order specifications. This is all right if the proposal is well detailed. However, if it just says "robotic welding system for welding backhoe buckets" then you will need to add some more terminology.

Again, try not to give the vendor a hassle about lead times unless it really is crucial. The vendors want to build the system as quickly as

143

possible so they can ship it to you and can collect their money. They typically do not quote delivery times that are any longer than necessary. Include the quoted delivery in your order, after verifying with the supplier that the quoted delivery is still valid, especially if the quote has been on your desk for 3 months. Vendor backlogs can change quickly.

Be sure to spell out all the critical components of the cell and all options you wish to purchase. More care should be taken here than when ordering other types of machines, since welding systems can be very unique. The design of the system, the features, and the options can be very specific to the function and purpose of the equipment. A machining center might typically come with a particular type of control. On your welding robot, however, there are a dozen power supplies to choose from, even more welding torches, all types and sizes of workpiece positioners, and various brands of PLC and other logic devices for safety and cell controls. Usually the vendor will quote what they have found works the best, but you may have reason to specify certain brands of components.

Here are the major components that should be well defined on your order form:
- Make and model of robot or machine
- Make and model of welding power source
- Make and model of wire feeder (a four-roll feeder is recommended for automated welding)
- Make and model of welding torch
- Style and/or description of positioners
- Accurate descriptions of optional equipment, such as servo tracks, two-position indexing shuttle tracks, touch sensing, through-arc seam tracking, etc.
- Installation services that will be provided
- Training courses that will be provided
- Start-up support—if any—which was a part of the negotiations

You should receive an order confirmation reiterating exactly what you have ordered, and what the vendor plans to start building. Now is the time for any last minute changes, before the supplier begins building custom components. If you back out of the deal, you may be responsible to pick up the cost of any custom engineering and building of custom products that may have taken place up to the time of order cancellation.

Along with the confirmation, be sure to procure a document describing the plant requirements. These may include:

- Special foundation specifications, especially for larger systems
- Power requirements, amperage ratings of all equipment, and whether isolated power lines or special earth grounds will be required
- Floor space required and minimum ceiling height required for clearance of large systems
- Whether water or compressed air is required
- Where and how the system will fit into your plant

Teamwork

The team should be heavily involved at this point. There will be work for everyone to do. The operators can begin preparing the general layout of the cell, where the materials will be stored, and where the finished weldments will flow out of the cell. These are the people that know how it has to be, so give them the ability to make decisions regarding the surrounding cell layout. Obviously, they may need input from those knowledgeable about power availability, lighting, and product flow, among other things.

The maintenance and plant engineering people must begin the plant preparations. Foundation requirements are usually quite general, except when overhead mounts or large system components are involved. When an 800-pound robot starts swinging through the air at 10 feet per second, a lot of stress is generated in the overhead support structure. The structure needs to be firmly anchored to a floor that can withstand such loading and still maintain the repeatability and accuracy of the robot itself.

If not directly addressed in the quote, you may request a common utility connection for electrical power, so you only need to run one line to the welding cell. Individual power inputs are required for the robot controller, the welder, some external equipment, and any other power-consuming options you may purchase. The cell will be much neater and easier to service if the lines all run to a common fused junction box, and installation will be quicker and easier.

Most systems can be ordered to run on 240- or 480-volt supplies. Amperage service for robots and other computer controlled machines is typically about 10 amps. Welding power sources require from 40 to 60 amp service. An average welding cell may require a 100-amp line to handle all of the equipment, but be sure the supplier sends you all of this information in time for you to prepare your facility.

Shop air may be needed for such things as automatic spatter cleaners, pneumatic shuttles, or air-operated fixtures. Water may be needed for water-cooled fixtures, although water recirculators are normally used for this purpose, as well as for the water-cooled welding torches.

The engineers may have some redesigning of the product to do in order to maximize the benefits of automated welding. The time to do that is before the robot hits the floor. Have a layout of the system on hand so you can tell where interferences will be. This is especially true of tool designers, if you are attempting to fit your own fixtures to the cell. If your machine must sit idle for 2 months while your tool designers finish the fixtures, your payback will suffer considerably. Tool designers must work closely with product design engineers in order to optimize the manufacturability of the components.

If you are ordering a robot cell, then by now you should have received hole patterns for the positioner face plates from the supplier. This will allow you to adapt your fixtures to positioning equipment in the cell. Be sure to mount the fixtures so that the working range of the robot is optimized. Take into account the motions of the positioner, so that the part will be presented to the robot in such a way as to allow the robot to reach as many welds as possible, with as ideal a torch angle as possible.

Management may have some prickly problems to deal with before installation. Do you need to create a new union classification for operators of automated welding equipment? Does the pay schedule need to be adjusted? By eliminating a welding bottleneck, a pileup may occur elsewhere in the plant. You need to anticipate if and where that may happen and formulate a contingency plan for when it does. This may even involve purchasing other new equipment to keep up with the welding cell, such as CNC brake presses or shears, machining centers, or CNC cutting tables. Remember that you are not just buying a machine; you are changing the way you do business.

Management must plan to give the operators authority to make important decisions regarding the daily operation of the system. This must be communicated plainly to them, so they will be encouraged to lend their own ideas to the project. Checks and balances are, of course, necessary. By giving operators control over their work environment, you may also be making them accountable for their actions. This removes a burden from middle management and allows decisions to be made much more rapidly and effectively.

Know the plan

A project schedule is a good idea. It can give you a somewhat realistic view of what is happening. Please, however, have mercy if your supplier misses it by a couple weeks. Unlike many other types of equipment, welding systems are typically quite unique. They are built from standard components into a cell customized for your application. Sometimes even the components are not standard and are specially engineered for you. This requires time, engineering, and a lot of thought, all of which are difficult to predict and all of which can mess up quoted delivery times. Obviously, if a project starts running many weeks or months behind schedule, something went seriously awry, and your complaints would likely be justified.

Important milestones might include (not necessarily in this order):
• The vendor ordering components for your system
• The vendor receiving those components
• The vendor beginning design and engineering of your system

- Design of welding fixtures
- Your approval of fixture design in order for the vendor to begin building the fixtures
- The beginning of the assembly of your system
- The completion of its assembly
- The first time the system is under power
- The beginning of part programming, if a robot cell
- A scheduled runoff
- Breakdown and shipment of the system
- Delivery and installation of the system to your facility
- Startup and debugging of system
- Production programming of the system
- Fine-tuning of system parameters and functions (ongoing)

Pay attention!

While your system is being built, pay attention to the project! Visit the vendor's factory as often as you can while your system is being built. Even after taking all the precautions possible, your supplier may have a slightly different idea of what you want and need than you do. Since you know that the squeaky wheel gets the oil, you may find that showing your face on a regular basis can keep the project moving right along.

Compare the progress you see with the previously discussed project schedule. If you see trends toward a shorter or longer lead time than expected, you can make the necessary adjustments back home and address these issues while you are standing there in the office of the person who quoted your delivery date. By visiting often, you will gradually become intimately familiar with your new acquaintance. Your welding system will be a part of you before the delivery even takes place, so you can hit the floor running.

The runoff

In marriage, it is not possible to have a trial run. In the welding world, however, you would be foolish to go on without one. Suppliers should include a way of proving that the system will do what it is intended to do, and this is especially true of custom welding equipment and robotics. I am not talking about a preorder demonstration, but of a bonafide production runoff on your system prior to shipment to you.

For hard automation, a trial run of ten parts or so is probably sufficient. Since the variables are minimized, you are somewhat well assured that if the system can run 10 parts, it can run 1,000. With flexible automation, which is intended to run wide varieties of parts, you should request that the vendor demonstrate a typical shop-floor production changeover to you. It may not be necessary actually to run several of each part but make sure you know that the system will perform as expected on the parts that were originally specified.

For robotics and other flexible automation, a runoff of 20 parts is probably sufficient to determine the functionality of the system, if the cycle times are short (a couple of minutes each). If you need to supply SPC-style statistics for your customer, then you may need a 300-part run from the welding supplier. Vendors will do this for you, but most will charge for it. Expect to pay for any runoffs that take more than a day or two.

If the cycle times are very long, then a runoff of one or two parts will tell you what you need to know. No one wants to stand and watch a 25-part runoff if the cycle time for each part is 3 hours. If you see a couple of ugly welds here and there, do not worry about it. If it is something that can be remedied by changing a few weld parameters, then rest assured that the robot will do the job. Don't waste a lot of time demanding an absolutely perfect part during the runoff. After all, you purchased the robot for its flexibility. You know that it is programmable, so you know that minor problems can be adjusted after installation. For very large, complex parts, you will fine-tune the programs for some time before you are completely satisfied with the results. This is something that should probably take place on your floor, not the vendor's.

The above descriptions of runoffs assume that you have purchased fixturing from the automation supplier. If you have decided to build your own fixturing, then do not expect a production runoff before shipment. The vendor's only responsibility in this case is to simply prove that the system functions as designed.

Honeymoon: Getting Off to a Good Start

Probably the most exciting part of such a project is the delivery, when all that you have thought about and discussed for the past few months suddenly materializes on your receiving dock. This is where the honeymoon begins. The days are filled with anticipation as you begin installing the system you have envisioned for so long. If you keep your feet on the ground, and remember how you got to this point, then you will be able to begin building a new way of life that will ensure success.

Training

Training is an obvious fundamental, but do not be fooled by those who tell you that 1 week is enough. Sure, most training classes for more complex machines like robots will last for 1 week, but that is only to get the future operators familiar with the protocol. I am not suggesting that more than 1 week of vendor-sponsored training is necessary; what I am saying is that the real learning takes place on your floor, after the system is installed and after the operators become responsible for making it tick on a day-by-day basis. You must allow adequate time for this critical activity.

One week of training is normally enough time to learn the basics. However, in addition to programming or basic machine functions, the trainees should have a thorough explanation of the hands-on daily operation of the system. A good trainer will focus on exactly what a person needs to know to keep parts running through the system hour by hour, and this type of practical experience is sometimes lacking in such courses. Understanding the shortcuts will enable your operators to shorten their learning curves and to utilize this new tool better.

Of course, hands-on training is the best approach, and most training classes emphasize a large majority of hands-on operation during training. This is especially useful if such training is performed on the very machine that you have purchased. Whether this on-system training takes place at your supplier's facility before shipment or on your own floor after installation, it will add to your lead time and delay the startup. However, it is well worth the extra time and effort to make it happen this way. Since automated welding systems are generally custom designed and often quite unique, this type of training is much more effective than the theoretical classroom counterpart.

Be sure to include maintenance training, especially with the more sophisticated machines. You simply cannot afford to have such an investment sit idle while some minor problem is solved. Proper training on not only maintenance, but also on preventive care, is worth the extra expense and will pay for itself quickly in higher up times and better productivity.

During training, you will be able to identify individuals with a particular knack for certain tasks. You may want to identify who your star programmer will be and who will best understand the care and repair of the system and then keep them involved. Some shops create new labor classifications for operators of automated equipment. This can be good incentive for employees to become more interested, involved, and committed to the project. Regardless of the approach, be sure to leave enough time, in addition to the formal training, for your people to grasp what has happened, what their new responsibilities are, and exactly how they will be functioning in the months or years to come.

Installation

Installation should mainly be provided by you, the end user. Most automation companies discuss "installation supervision," which means that they will provide a technician to help direct your people, but it is your responsibility to provide labor and machinery to unload the equipment from the truck, set it up in your shop, fasten it to the floor, run utility connections, and do various wiring and other types of connections.

If the supplier offers to provide more support than that, take it.

Some things are done better by them anyway, such as certain wiring connections and adjustments. You may also consider hiring them to provide complete installation services, especially if you are short-handed. Obviously, this will increase your cost, so it is important to define precisely what is provided.

If the system was functioning properly in the vendor's factory prior to shipment, installation on your floor should take only a day or two for typical systems, but up to a week or more for large or complex systems. If programs were written for a runoff, then they must be fine-tuned after installation, since the system will most likely not go back together exactly as it was. Variations in the surface of your floor are enough to throw off a welding program significantly after the system is bolted down.

Typically, it is best to allow your own operators to program or otherwise prepare the system for production, in order to get them started through the learning curve. You may also hire the supplier to write additional programs or prepare the system to run additional part numbers after the system is installed. Either way, you must allow for this in your scheduling and be aware that your startup will be delayed that much longer.

Making the transition

Rather than demanding 110% production from your system the day after installation, an easy transition into automated welding will accomplish several worthwhile goals:

1. Operators will have a chance to get through the learning curve effectively, becoming true experts in automated welding. If the process is rushed, shortcuts will be taken that will compromise the future reliability and productivity of the cell. Success depends on the operators becoming intimate with the function and operation of the machine and on their knowledge of its quirks and its likes and its dislikes.

2. If the welding system will be eliminating welding bottlenecks, then you could be creating bottlenecks in other plant processes. With a sudden and explosive startup, the demands placed on other departments can be shocking, causing all sorts of chaos. A more

gradual startup allows the whole plant to become accustomed to this new way of life, especially if the equipment is displacing a large volume of work.

3. To expect full production immediately is usually expecting too much. If you depend on the machine to weld all of your product too soon after startup, then you may pay dearly when it goes down due to a lack of operator knowledge. Until the operators are comfortably through the learning curve, there will be times when the system just sits there waiting for a command that the operator just cannot remember. It may take an hour on the phone to figure out the problem. Meanwhile, you are an hour without production. It is better to avoid costly downtime penalties by making sure that the operators are comfortably knowledgeable regarding all the intricacies and nuances of the system's functions and operation, before scheduling all of your vital production through the machine.

4. Scheduling difficulties will ensue after the installation. It normally takes some time before products begin to flow as planned through the cell. Give it some time in order to avoid running out of parts or causing another department to run out of parts.

Establishing a routine

Routings, part volumes, product flow, order quantities, and other production schedules may all be disrupted by the increased performance of your new welding system. To address these factors ahead of time will get you closer to the goal of establishing a routine. This will allow you to predict accurately the production available from the machine, the consumption of consumable parts and materials, and the amount of labor and other resources necessary to keep things flowing smoothly.

The Marriage: Making It Last a Lifetime

You must understand what this new way of life is really all about. You should work at keeping this romance alive and well. Successful users of automation do. How would you otherwise explain those that own multiple welding systems? How about those robot users with a dozen robots in their factory? These users have found the key to true happiness with welding automation, and these users are the ones who are leading and will continue to lead the way in all sorts of accomplishments, simply because they want to be the best at what they do.

A new way of life

After all the glamour and glitz is forgotten, a machine is still a machine. If it is a robot that everyone is excited about, rest assured that the dust will settle, and the robot will become just another tool by which you can improve your operations. This is my goal: to get you so well accustomed to automation in your welding department, that it will be as natural for you to buy an automated system as it now is for you to buy a welding power supply and wire feeder.

Here is my possibly futile attempt at prophecy. More than 20 years ago, CNC machining centers were the latest rage. Most people thought they could not afford them, and the rest were simply afraid of the new technology. As time went on, and the industry matured, people became convinced of what these new computerized machines could do for them. Production rates tripled, quality increased, inventory was reduced, and profits increased. I do not need to describe the multibillion dollar industry that has ensued.

Here is the prophecy: I believe robots and other computerized welding equipment will become the commodity that machine tools have become. At present, a majority of welding robots are sold in cells. The day will come when, like machine tools, many automated welding machines will be purchased in crates by end users. These end users will become system integrators, because the technology will no longer be intimidating or out of reach for companies like yours.

How do we, in the welding industry, arrive at this point? By believing in and embracing welding technology for what it is: the only method available of changing the way we have done welding for the past 30 years. New advances are being made daily in welding technology. Even welding power sources are now computerized—

some are even artificially intelligent—and are capable of providing unparalleled weld quality and ease of operation. Welding is the last of the major manufacturing processes to receive attention from the technology revolution. When you consider that welding is one of the most labor intensive manufacturing processes, it stands to reason that the returns from welding automation should surpass those from most other capital investments that you can make.

Tracking results

If you cannot show that your new investment is saving money, improving quality, or reducing inventory, then chances are it will be your last such endeavor. Our goal, then, is first to insure success, then to build on that success to justify further automation. If the volume of work is available, most successful installations in the field are followed by many more. This is a testimony to the attractiveness of such an investment.

Good record keeping will allow you to track your progress properly and prove to the world that you made the right choice. This obviously must take place prior to or during the events discussed in previous chapters. Like reading a map you must first know where you are, before you can determine where you are going. Detailed information regarding your current costs will make it much easier to justify an investment in welding automation, especially if you include not only base labor rates but also current levels of consumable usage, scrap and rework levels, indirect labor time, costs for subsequent cleaning operations, and all the other factors we discussed earlier.

Comparing these old figures with the new rates resulting from your automated welding installation will provide you with much valuable information.

1. You will be able to verify your ROI estimates, which will probably be more attractive than you had initially calculated.

2. You will learn on what items your money was well spent and which purchases were frivolous or unnecessary. You will also discover what you should have purchased. This is extremely valuable information to have when you start your next welding project.

3. You will have the ammunition to convince management that

welding automation is achievable and necessary and that it is
probably the best investment they have made in a long time.

4. It will make it much easier to conceive, justify, and install your
next automated welding system.

Summary

The job is just beginning when you place the order. Any supplier
out there can sell a machine and accept an order. What happens from
that point on determines your degree of success or failure. Follow
these guidelines to remove some of the risk and to increase the
profitability of the system you purchase.

A. The courtship: excitement and nervousness.
   1. The engagement
      a. It is extremely important to define the project accurately so
        that the purchase order will describe in detail what you, the
        buyer, want to purchase.
      b. You should receive an order confirmation describing what
        the supplier thinks you are buying. Check it out carefully.
   2. Teamwork
      a. Each team member should have specific responsibilities.
      b. Maintenance and plant engineering need to get the facilities
        ready, including foundation and utilities.
      c. Product engineering may need to do some redesign to make
        components more automation friendly.
      d. Management concerns:
        (1) Are we creating a bottleneck elsewhere?
        (2) Do we need more equipment in order to keep up or to
            produce parts with adequate tolerances for automation?
        (3) Do personnel changes need to be made, such as a new
            pay scale or new labor classification?
   3. Know the plan
      a. Ask for an accurate project schedule from the supplier. This
        will help expedite your preparations.

4. The squeaky wheel
   a. Visit the vendor as often as possible during production of your welding system.
   b. Use these opportunities to become familiar with your new equipment, and to fine-tune the delivery schedule.
5. The runoff
   a. Insist on a production runoff.
   b. Be reasonable. Have realistic expectations, realizing that it will take a few days or weeks to get the machine purring on your shop floor.

B. The honeymoon: getting off to a good start
   1. Training
      a. One week is usually enough, but the real training takes place after installation.
      b. Hands-on is the best approach (lots of it).
      c. Maintenance training is also important.
      d. You must leave enough time, in addition to formal training, for your operators to feel comfortable with the system and to learn to use it properly.
   2. Installation
      a. Usually, installation will be mainly provided by you.
      b. Programs will most likely need fine-tuning after installation, so allow time.
   3. Make the transition to automated welding an easy one because:
      a. Operators need a chance to get through the learning curve effectively.
      b. You will need time to figure out how to deal with other bottlenecks that may be created.
      c. Lack of preparation could cause excessive downtime in the early stages of the game.
      d. It takes some time to readjust scheduling according to the new and improved production rates.
   4. Establishing a routine
      a. Try to anticipate the new schedules and routings. This will

help you toward your ultimate goal: to get back into a routine.

C. The marriage: making it last a lifetime.

   1. A new way of life

      a. Your new automated system is just another tool for improving the way you do business. Automating your welding operations should become as easy and natural as purchasing any other type of machine.

      b. The prophecy: robots will (at least should) become the commodity that machine tools have become. End users will purchase fewer of them from integrators and do the integration themselves. As an industry, we are not there yet, but some companies have arrived at that point.

   2. Tracking results

      a. Good record keeping before the installation is a must.

      b. Good record keeping after the installation is a must.

      c. This will allow you to:

         (1) verify ROI estimates

         (2) determine if it is necessary to do something differently next time

         (3) convince management that it was the correct choice

         (4) purchase the next system more easily

8. Questions and Answers

Below are some common questions that are raised during discussions regarding implementation of welding automation. You may find some of your own concerns being identified in this section.

Q: What if I have low part volumes?

A: Low part volumes are much less challenging now than in the recent past. It is important, however, that these low-volume parts repeat on a regular basis, be it daily, weekly, monthly, or longer intervals. The ability of a robot, for instance, to recognize a part based on some type of digital input is commonplace today. This implies that lot sizes of one may be fixtured on pallets and identified automatically by the robot or by some simple operator input, such as a thumbwheel. For robotic welding and for hard automation, flexible fixturing can allow several parts to be welded on a single fixture. This eliminates changeovers and greatly reduces fixture costs.

Q: My part fitup is not consistent enough for automatic welding. What can I do?

A: Part fitup is one of the toughest problems to overcome in automated welding. It is not insurmountable, however. Reliable technologies exist that enable the machine to locate weld joints. These include touch sensing, through-arc seam tracking, laser tracking, vision, and others.

The first recommendation is to fix your fitup problems. This will also improve your overall part quality, so it can be a worthwhile endeavor in itself. The above options, some quite costly, can only

159

provide limited corrections. One problem that is very difficult for any tracking system to handle is gaps in the weld joint. Again, the best solution is to improve your parts.

Q: How much time does it take for the whole implementation process, from the time I decide to automate to the time the machine is running productively?

A: It is virtually impossible to predict the real development schedule of such a project, since the variables are so numerous. These variables include:

- What type of automation are you purchasing?
- To what degree are your components ready for automated welding?
- How sophisticated is the technology?
- Is this a new plant startup, or are you automating an existing product?
- What types of constraints are present (time, money, availability of operators for training, "trainability" of your employees, availability of materials, constraints on individuals due to other projects)?
- What type of resources can you make available specifically for this project?

With such a disclaimer, you may think that *any* guess at a project schedule would be futile, but there are certain assumptions that can be made. A simple robot project can take from 8 to 12 weeks (all estimates are subject, of course, to present backlogs). More complex projects may take as long as 6 to 8 months. One successful builder of hard automation and robotics cells sometimes has backlogs as long as 9 months. Their customers obviously think it is worth the wait. So many factors affect the lead time of such equipment, that it is best to discuss it in detail with those who are providing your quotations.

Q: Our plant would like to become ISO 9000 certified. Will welding robots help our cause?

A: First of all, some readers may not fully understand just what ISO 9000 is all about. ISO 9000 refers to a standard for quality systems that has been adopted by the European Economic Community and is being sought after by many companies here in the U.S. Many manufacturers expect that they will not be able to do business in Europe without the certification and that, eventually, the same will occur in this country. Rather than a set of quality guidelines, ISO 9000 consists more of requirements for implementation of a system; not a canned system dictated by the organization, but a system that you as a manufacturer are somewhat free to design for yourself. The important factor is that your system is workable and that you can prove to the auditors that your employees are well-versed in how your system works and that everyone knows where to go for the right answers. In becoming ISO 9000 certified, it is more important to know where to find answers—and to have those answers available to all employees—than it is to memorize a bunch of slogans or quality lingo.

On whether automation will help your certification efforts, the answer is both "yes" and "no". Welding automation gives users the ability to remove many of the variables of manual welding. By utilizing a computerized welding program, variability is diminished or eliminated and can be monitored and recorded as well. These are all important aspects of this quality program. However, the main thrust behind ISO 9000 is not to provide manufacturing standards but to cause companies to conform to their own standards. If you write a quality manual and have thorough quality procedures, then certification depends on how closely your operations conform to your own documentation.

Automation cannot help you become certified, but it can remove much of the variability in your welding processes. This provides consistency and measurability, which are required in order for you to gain control of your welding processes.

Q: I've done my time studies, and it looks like a robot would be no faster than my present operations. Why automate?

A: There are indeed applications that will not justify themselves on the surface. We have discussed, however, variables other than mere labor savings; variables that are typically not used in justification schemes. You must pull out all the stops in order to gain a realistic view of what your true savings will be. I am not suggesting that you alter facts. I am requiring that you consider how much you spend today to run your business, then compare that with how much it costs after your automation gets up and running. Take into account every conceivable expense, and you will see a significant difference. Also consider whether higher quality and better consistency will allow you to increase your customer base or whether such goals are necessary to become ISO 9000 certified. Just one of these factors alone can adequately justify even quite expensive equipment. Very often, solving one problem provides the solution for several others simultaneously.

Q: I heard about a robot system that failed miserably. I do not want that cloud hanging over my head.

A: The robotic and automated welding industry has matured. Most of the stories you hear of lemons happened years ago, while the industry was still young. As in any other business, some systems are still installed that never seem to function as planned. However, this is often due to either unrealistic or improper expectations from the end user or poor definition of the project up front. Do not let rumors extinguish your desire to pursue automation. No matter what you have heard, the benefits are real and achievable, provided you have a proper application. Since welding is one of the most labor intensive processes, it makes sense to at least investigate. The ROI you may achieve can be quite attractive.

Q: How can I afford a robotic welding system? I'm just a tiny company with a tinier budget.

A: First of all, your application may not require a robot. Automated welding systems are available that often cost a fraction of a robotic welding system. You need first to define the correct type of automation for your needs. Second, you need to ask yourself some pointed questions.

1. "Are my competitors automating?" If they are, you may be in for quite a fight unless you do something about it. If they are not, now is your chance to get a jump on them; this could give you an advantage that will be difficult for them to overcome.

2. "Is there a good supply of welding labor out there, and will there be in the years to come?" Many are predicting a shortage of skilled trade workers in the future. You may need the extra productivity that comes from automation just to continue at your present levels.

3. "Do I want to remain a tiny company?" One of the surest ways to grow and succeed is to automate. There are enough success stories around to prove that this is the case. Continuous improvement should be the intent, which is typically followed by higher quality, lower costs, less inventory, and higher sales. This is a prescription for growth.

Q: I am afraid our welders will revolt if they find out we are considering automation.

A: This demonstrates the importance of communication and making the shop floor welders a part of the automation team. The only reason people will rebel is ignorance. While I must admit that sometimes automation causes loss of jobs, the overwhelming majority of cases prove that this is not at all the norm. Most often, companies grow and prosper as a result of modernization and automation as they leave their competitors behind.

One of the first jobs you will have as a champion of welding automation is to sell your operators on the idea. Do not make the mistake of hiding it until the machine arrives on the dock. That simply undermines their faith in you. Such a project can be used as a method to unite your work force and to instill in them a sense of pride, accomplishment, and worth, since they really are the ones that best understand the processes that you will be automating. Communicate with them and, more importantly, listen to their ideas.

Q: What if our parts vendors cannot provide components to meet the requirements for automated welding?

A: How to work effectively with vendors has been the subject of volumes of books, manuals, and articles. To address it here adequately would be impossible. One approach that has been taken by existing users of robots and automation is simply to require that vendors automate, whether that involves welding, flame cutting, machining, stamping, bending, punching, or any other manufacturing process that can be automated.

It is not realistic to simply demand that vendors provide parts to tighter tolerances when they have no means of manufacturing such parts. What will really happen is this: their production procedures do not change, they simply begin 100% inspection. This may sound great on the surface, since it guarantees good parts for you. However, your vendor must still make a profit. To cover the extra costs of inspection, and the higher cost of scrap and/or rework, they will either raise prices or settle for a lower profit margin. This is not a healthy environment in which to do business, because you should rely on healthy, stable vendors just as you rely on stable, profitable customers for your livelihood.

By encouraging them to automate, you can accomplish several goals simultaneously: your incoming parts will be more consistent, your supplier may be able to charge less by passing some of the cost savings on to you, and your supplier may become more prosperous and able to afford even more modernization and better technology. I am sure that you can list many others.

9. Robotic System Layouts

This chapter simply consists of layouts of various robotic welding systems. The possible configurations of such work cells are as diverse as the types of weldments to be welded. Perhaps you will see something here that could work for your particular application. This is just a small sampling of the ways in which standard system components can be fit together into a custom welding cell. Remember that an experienced automation engineer must assist you in your plans, in order to specify the equipment that will best meet your needs and to provide the best value.

Figures 9-1 through 9-4 describe how a robot works and moves, with explanations regarding the robot's working range. These drawings are to make you familiar with how a robot operates so that you will better understand the complete systems which follow.

In general, small, simple parts that require no positioning during welding are good candidates for such cells as shown in Figures 9-7 through 9-10. These all include flat surfaces to which you can mount welding fixtures, and they also include a means to weld internally to the load/unload cycle.

More complex weldments that require simple positioning may be welded in cells like those in Figures 9-5, 9-6, and 9-11. These include one additional axis of positioning that aids in reaching more welds, and in positioning the welds properly in relation to gravity.

The most complex positioning motions are accomplished in cells such as those shown in Figures 9-6, 9-14, and 9-18. As work cells become more complex, they not only become more expensive, but the learning curve is lengthened somewhat. It simply takes more time, more money, and more effort to get such systems on line, but the

165

financial returns can be fantastic. Look at the layouts, and get some
ideas for your own particular applications.

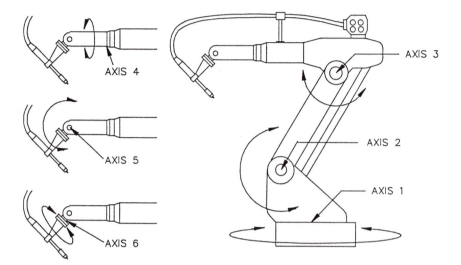

Figure 9-1. The working axes of a six-axis robot.

The six-axis robot is the most common configuration for welding
today. Although five-axis robots are available, they are quite limited
as compared to those with six axes. A five-axis robot, however, may be
adequate, depending on your specific application. The advantage to
these less flexible robot arms is that the cost is usually less.

The robot shown is quite typical in design, and consists of a base
rotation motion (Axis 1), and two degrees of freedom in the arm (Axes
2 and 3). These three axes are mostly responsible for moving the
welding torch from weld to weld and also for the speed at which the
robot makes these motions. When various robot companies vie for the
title of "fastest robot," it is usually a result of these three axes being
driven by larger, more powerful motors.

Axis 4 usually takes the form of a concentric rotating upper arm,
although various designs exist. The fourth axis is the one that is
missing from most five-axis robots, and it allows the user to approach
welds at angles not possible with a five-axis arm. Axes 5 and 6 are the

wrist axes, which allow bending and rotating of the robotic welding torch or other end effector. Axes 4, 5, and 6 are mostly responsible for achieving and maintaining the proper torch angle during welding. These three axes typically do not contribute as much to sheer positioning speed as the first three axes do, but they are crucial for accuracy and for other functions such as weaving.

Some robots have the capability to do "wrist weaving." This, in effect, allows a much faster weaving speed than standard weave functions, since standard weaves are accomplished by oscillating the entire robot arm. By relying on the wrist for the weave motion, wrist weaving can produce weave speeds of up to five oscillations per second or faster.

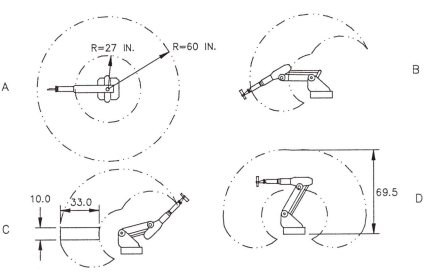

Figure 9-2. The robot's working range.

Most modern robots are quite flexible, as shown in this drawing. This robot can reach below its own base, can bend over backwards, and rotates 360 degrees. With six axes, there is almost no position that these robots cannot get to. View A shows the robot from a plan view, with the minimum and maximum working ranges identified. View B shows a side view of the robot, with its arm and wrist fully extended, while view C illustrates the flexibility of the robot as it bends backwards. View D shows the complete working range of the robot. Since the robot can rotate about its base 360 degrees (some cannot rotate a full 360 degrees), the work range shown should actually be viewed as a "donut" shape surrounding the robot.

The working range shown is measured at the center pivot of the wrist. Even though the welding torch actually reaches further, that is not necessarily all usable working range. When the robot's arm and wrist are both fully extended, as shown in view B, there is very little movement left in the robot axes. This makes it very difficult, if not impossible, to achieve proper torch angles for welding, and torch angle is a critical welding variable for good weld quality. In the same way, if the robot tries to weld at the innermost part of its working range, it will have no useful torch angle left. This is why it is important to understand exactly what a drawing of a robot's working range tells you.

In view A, the plan view of the working range shows that the robot has a maximum reach of 60 in. This statistic shows up in all of the robot literature available today, because it is obviously a valuable piece of information to have. The inner radius of 27 in. is meant to represent the minimum effective working radius of the robot. As you can see in view B, the robot is actually capable of almost touching its own base, but this is not part of the effective working range. The usable working range is that volume in space in which welding fixtures and weldments could be placed where the robot could effectively reach the welds. In view C, assume that there is welding necessary on the side of the box facing away from the robot. Without a positioner to bring the weld to the robot, the robot would not be able to reach it. Even though the welding torch tip reaches beyond the edge of the box, as seen in view A, you must bend the robot wrist in order to achieve good torch angle. This uses up working range, so you will not reach all the welds, even though it appears on paper that it would work.

Another limitation to the working range of the robot concerns its shape: a donut. Most weldments have a generally square or rectangular profile, which does not fit well into this circular hemisphere.

Thus, it is very important for you to understand fully how the robot arm functions and how it is manipulated into various positions. Do not assume that all of the working range shown in a brochure is usable. Some robotic companies actually show their robot arms bent at 45 degrees in their layouts, to assure that a good wrist angle can be achieved during welding. You will notice that in the side views, the working range nearly touches the robot base itself, but, in the plan view, the inner working range is several inches from the robot. This is so that a truer work envelope is presented, as illustrated by the placement of the box in view C. The part of the working range that coincides with the inside edge of this box is where the 27 in. inner radius comes from.

View D shows the maximum height of the robot's work range. Although this figure is rarely needed for a floor-mounted robot (any fixtures located in this area would be quite difficult to load and unload) it is important when considering an inverted mount for the robot. Figure 9-3 discusses inverted robots in more detail.

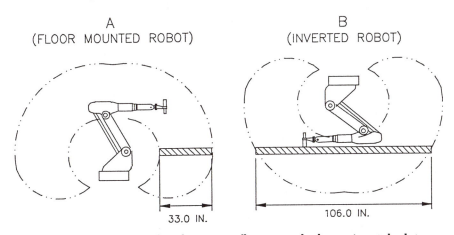

A
(FLOOR MOUNTED ROBOT)

B
(INVERTED ROBOT)

33.0 IN.

106.0 IN.

Figure 9-3. Comparison of working range: floor-mounted robot vs. inverted robot.

View A shows the effective working width of the robot in question. The space taken up by the body of the robot itself is, of course, not available for fixturing parts or for locating positioners. If a part is wider than the indicated 33 in., it may be necessary to manipulate the part during welding or to move the robot along a track in order to reach all of the welds.

A solution to this problem is found in inverting the robot and mounting it upside down to some type of structure. As shown, this increases the effective reach of the robot to 106 in. The useless working range directly above the floor-mounted robot is made available by inverting the robot, so much larger parts can be welded without moving the robot. For this reason, many robot systems are mounted in the inverted position.

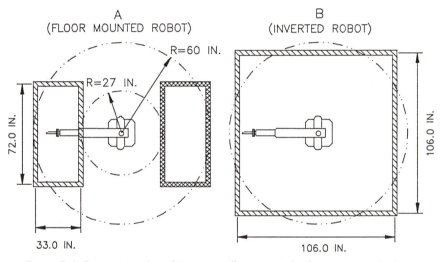

Figure 9-4. Comparison of working range: floor-mounted robot vs. inverted robot.

This drawing shows the same robots depicted in Figure 9-3 but from a plan view. The heavy black boxes represent the effective working range of the robots. Remember that we are trying to fit a square peg into a round hole, so the corners of the square boxes lie outside the robot's work range. This is one of the challenges in applying a particular robot to a certain job. Most fixtures and positioners are based on some type of square or rectangular framework, but you must determine whether the actual weldment will fit within the robot's reach.

The floor-mounted robot, again, has a much smaller usable range—a little over 60 ft². The advantage with an upright robot, however, is that you can put a second or even third workstation on the floor, around the robot. The inverted robot, on the other hand, can cover about 78 ft². This work area is normally used for only one workstation, since placing two stations under the robot would present a danger to the operator—it would be very difficult to safeguard. For this reason, inverted robots are usually used in conjunction with a positioner on a rotating base, such as those shown in Figures 9-14 and 9-16, or on an overhead track.

With most modern robots, no special changes are made in order to invert a robot, but there is a common problem encountered with upside-down robots, especially during teaching. When you press the "up" button, the robot will move down, since it is hanging upside down. Some companies allow the user to change machine parameters to fix this annoying problem. Others may not, but the operator will become accustomed to it in time.

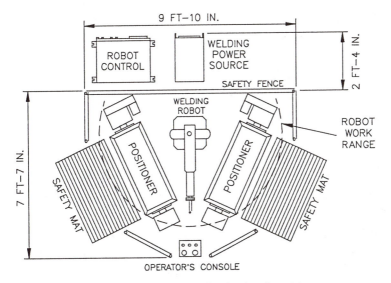

Figure 9-5. Robotic welding cell with two head/tailstock positioners.

This cell (Figure 9-5, page 171) consists of a welding robot working in conjunction with two single-axis, head/tailstock (HTS) positioners. These positioners can be either servo-controlled (infinitely programmable) or indexing units. Indexing HTS positioners may index every 90 degrees, or every 30, or almost any combination you desire.

Of course, there are two stations, so that the loading and unloading of parts can be accomplished internally to the welding cycle. A disadvantage to arranging two stations on opposing sides of the cell, as shown, is that the operator must do a lot of walking, from one side to the other. If the total cycle time is short, then he will make this trip perhaps hundreds of times per day. If this is the case, there are better solutions, for instance, those shown in Figure 9-12. If the robot must wait for the operator, then the productivity of the cell will suffer.

Each HTS positioner is guarded by a safety mat. If the operator steps on the safety mat while the robot is at that station welding, all equipment stops immediately to prevent harm to the operator (or anyone else, for that matter). The robot controls keep track of which station is the active station and which safety mat needs to be monitored. The cell is surrounded by a safety fence, to prevent others from inadvertently walking into the work range of the robot. It is especially important to protect those who are not familiar with the cell, since they may not understand how fast the robot can move or when it is about to move.

HTS positioners are generally required for more simple weldments, but it is surprising how versatile they can be. If your weldment has many welds at various angles, but all at 90 degrees to each other, then a single-axis positioner may get all of those welds in either the flat or horizontal position. This is usually enough to provide sufficient weld quality for all but the most strict requirements. If the angles of the welds prohibit effective positioning with the single axis, then an orbital positioner, as shown in Figure 9-6, may be necessary.

The following cell (Figure 9-6) utilizes the more versatile orbital

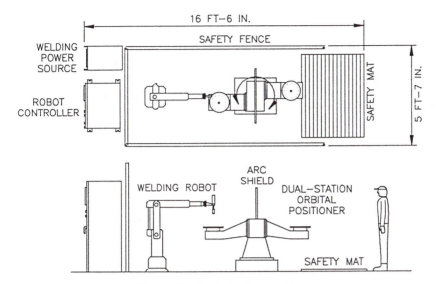

Figure 9-6. Robotic welding cell with dual orbital positioner.

positioner consists of a rotating arm and a platter on the end of that arm that also rotates. This provides two axes of motion per side. This particular positioner consists of two orbitals mounted on a rotating base. This minimizes the amount of walking the operator must do while loading and unloading parts.

A safety mat is also used here, although for a slightly different purpose. The biggest danger with this system is that a person's clothing or anatomy might get caught on the positioner while it is exchanging workstations. Once these positioners start rotating, it takes a lot of force to stop them—much more than a human body can muster. Thus, the safety mat's main purpose is to keep people out of the cell during station exchange. After the positioner is finished rotating, then the operator is free to step on the mat, approach the positioner, and begin unloading completed parts. The robot will then proceed to weld on the station that is facing it.

The arc shield at the center of the positioner protects the operator from arc glare while he works at the positioner. The safety fence surrounding the cell typically blocks arc rays, too. This is done

either by building the fence of a solid material like sheet metal, or—
more commonly—to construct it out of transparent plastic sheeting
specifically designed to block the harmful arc rays. This material is
available from almost any welding distributor in sheets of various
sizes. They are available in many colors, and the aesthetic quality of
these see-through arc shields is commonly preferred.

The robot controller and welder are shown behind the cell, but,
for convenience, the controller may be located closer to the front of
the cell, where the operator has better access to it. It is often
necessary, especially when working with frequent changeovers and
small lot sizes, for the operator to access the robot controller. An
alternative to locating it near the front of the cell is to build a remote
stand that is designed with certain control capabilities. The teach
pendant typically provides the major link to the control system, so
mounting this on a stand near the operator may allow you to leave the
controller itself hidden in the dark, out of the way. Besides, even
though they are designed for use in an industrial atmosphere, the
controller is safer if kept away from the sparks and smoke.

This cell (Figure 9-7) is specifically designed for small, simple

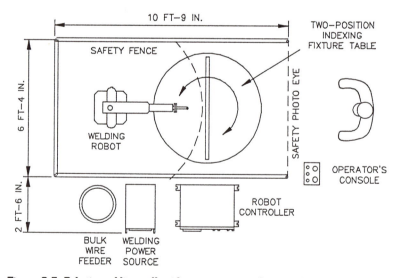

Figure 9-7. Robotic welding cell with two-position indexing table.

weldments that require no positioning during welding. The flat indexing table acts only as a station exchanger. The alternative is simply to locate two small, flat tables within the working range of the robot, but then you must deal with the increased mileage your operator will travel per day. Since the products welded in such a cell tend to be very simple and quick to weld, this means that the operator will be hustling to keep up even with the indexer helping him out. These indexing tables are quite inexpensive and usually well worth their cost.

If the parts to be welded are small, and the fixtures to locate them for welding are inexpensive, then multiple fixtures can be mounted on each side of the table. Say your cycle time for one part is 8 seconds. If you mount only one fixture per side, the robot will take 8 seconds to weld it, then about 3 or 4 seconds to index the table. Indexing time alone is half of your total cycle time. Floor-to-floor cycle time is *12 seconds per part.*

Now, if you mount four identical fixtures per side, then the total welding cycle time is (4) X (8 seconds) or 32 seconds. Add 4 seconds in order to index the table around, and your total cycle time for four parts is 36 seconds, or *9 seconds per part.* By using duplicate fixtures, you have decreased your cycle time by 25%—a significant productivity boost.

A safety photo eye is used in this cell, rather than a safety mat. Typically, these two can be used interchangeably. A photo eye tends to be a little more thorough, since it can guard the entire opening to the cell, unlike a mat, which can slide on the floor. In the system shown, with a rotating station exchanger, the photo eye prohibits entering the cell only while the table is indexing. After indexing, the photo eye is deactivated, so the operator can enter the cell and reload parts. As far as the robot controller is concerned, the photo eye is identical to a mat: both provide a simple digital input into the safety circuit of the robot.

Also shown is a bulk wire dispenser. Bulk wire can save you money in a lot of ways. Most obviously, operators will change a 750-

pound spool of bulk wire much less often than a standard 60-pound spool. If an operator takes 10 minutes to change a 60-pound spool, then you can see how the time spent on this unproductive task can add up. In addition, bulk wire is simply less expensive to purchase than on smaller spools.

The bulk spool is located outside the cell, for easy access. The wire itself can be run through conduit to the wire feeder, which is normally mounted on the robot arm itself. Various plastic pulleys and accessories are available to help route the wire into the wire feeder. Bulk wire dispensers are recommended on most robotic welding systems, except where the wire type is changed often, say, from solid wire to cored wire. This is more common in job-shop atmospheres than in large part-volume operations.

The operator's console is located within easy reach of the operator. This usually consists of an emergency-stop button, cycle start button (perhaps dual palm buttons, for safety), a program stop button, and perhaps a restart button. Other controls, of course, may find their way onto the operator's console.

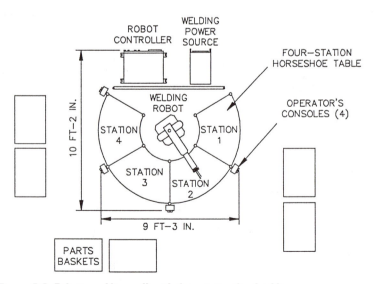

Figure 9-8. Robotic welding cell with four-station fixed table.

This horseshoe table (Figure 9-8) can provide a lot of versatility, especially where a variety of parts are being welded. The most common advantage to such an arrangement is simply that four different assemblies can be welded simultaneously. Obviously, parts cannot be manipulated during welding, since fixtures are mounted rigidly to a flat table.

Each of the four stations has its own "cycle start" button. Each of these buttons is tied into a hard input into the robot controller. The main control program of the robot is constantly looking at these four inputs, waiting for any of them to be turned on. When one of the "cycle start" buttons is pushed by the operator—say station 2—the robot will enact a subroutine that will call the welding program coinciding with the fixture presently mounted at station 2. The robot automatically resets the "cycle start" signal at station 2 to off and commences welding at station 2. If the robot finishes welding before receiving another input, then it simply returns home to wait for the next command.

If, on the other hand, the robot is still welding when the operator has finished loading another fixture, then the operator is free to push the "cycle start" button at that station, say station 4. Although the robot is still welding at station 2, the "cycle start" command will "latch" on. This means that it will stay activated until the robot finishes at station 2 and returns home. The robot then immediately sees that station 4 is activated and begins to weld there. If the operator can keep slightly ahead of the robot, then the robot will never sit idle. By maximizing the robot's working cycle in this manner, you are maximizing the effectiveness of your investment.

Another advantage to this horseshoe arrangement is that the robot can be welding at two of the stations—one station being welded while the other is being unloaded and reloaded—while someone changes fixtures on the other two stations. This allows some of the setup time also to be internal to the welding time. If setups and fixture changeovers are killing your productivity, then such a multistation system may greatly benefit you.

A third advantage is that the operator has less walking to do than if the welding stations were physically opposed to each other. This also provides much more space to place parts bins, shelves, or finished parts conveyors in and around the welding cell.

Since people cannot walk through the horseshoe table, the only safety fencing needed is behind the robot itself. Additional safety can be provided with automatically closing shields in front of each individual workstation. These are usually made of a flexible, durable material like leather, and are automatically closed when a particular station's "cycle start" button is pushed. When the curtain is closed, this indicates to the operator that the robot is in that particular station, and it also shields workers from the arc glare.

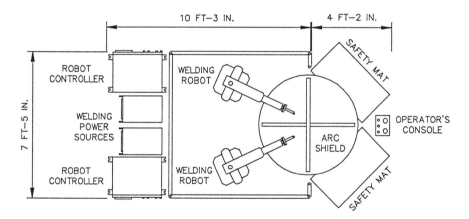

Figure 9-9. Robotic welding cell with two robots and four-position indexing table.

Since a robot usually has only one arc on at a time, the amount of weld metal it can deposit—although typically much faster than a person—can be limited. If very high cycle times are required due to high part volume requirements, then two robots welding simultaneously on the same part can greatly increase productivity.

Although the two robots could well be working on the same part, I have shown the two robots actually working at two different stations, out of a total of four work areas on the rotating table. In this scenario, say your weldment requires six welds. Robot 1 is to be responsible for welds A, B, and C, and robot 2 will do welds X, Y, and Z. The operator (maybe two operators) loads the fixture and presses the "cycle start" button. The table indexes, and robot 1 completes welds A, B, and C. Since the fixture that was previously at robot 1 is now at robot 2, robot 2 can complete welds X, Y, and Z concurrently with robot 1. Each index of the table provides a complete weldment but at twice the productivity than would be possible from a single robot.

To illustrate the amazing flexibility of robots, the parts at each station can even be different from each other. At the first index, robot 1 may make the appropriate welds on station 1, while robot 2 welds at station 2. After indexing, robot 1 can call a different welding program automatically and weld the first few welds of whatever part is in station 4, while robot 2 finishes the welding of the part that is on station 1. After indexing again, robot 1 can again begin welding a different part at station 3, while robot 2 finishes what robot 1 started on station 2. Four different part numbers can be welded, one after another, even though the welding is being shared between two robots.

Usually, some type of crosstalk is required when working two robots within the same cell. If robot 1 finishes before robot 2, then it must wait until robot 2 is also finished before indexing the table. Simple communication between the robots via digital inputs and outputs is all that is needed to prevent costly crashes.

Most robots require a separate controller for each robot. This is why two controllers are shown behind the cell. Of course, each robot must also have its own welding power source. This increases the cost of the system, of course, but it is not twice the price of a single-robot system. Since the two robots share the same positioner, safety equipment, floor space, and fixtures, the cost for a two-robot system as shown may be only 60% higher than that of a single-robot system, yet the productivity is twice as high.

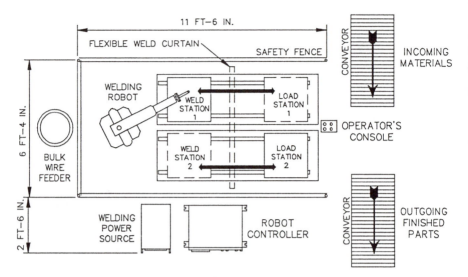

Figure 9-10. Robotic welding cell with two indexing shuttles.

Indexing shuttles are simply another means of exchanging parts between the load/unload area and the working area of the robot. In this example, two shuttles are used in order to keep the robot continuously occupied. While one shuttle is located at the robot's work area, the other is being reloaded. As a shuttle indexes toward the robot, it passes through a row of overlapping, flexible weld curtains. These are strips of filtering plastic about a foot wide, hanging straight down, allowing objects to pass through them but still shielding the operator from arc glare. The safety fence also helps protect people from the robot. Not shown is a safety-interlocked gate, that allows access to the robot for programming and maintenance purposes.

The shuttles themselves are simple flat carriages with a surface to which welding fixtures are mounted. Each of these two carriages is mounted to a pair of rails, allowing the carriage to slide smoothly and easily. Pillow blocks or similar bearing and shaft arrangements can be used. The carriage motion is provided by long pneumatic cylinders, usually capable of stopping at only two positions—one at each end of the shuttle track. Motion can also be accomplished by using ball screws or rack-and-pinion systems, which may allow further flexibility by providing more than two stopping points for the carriages. Pneumatic exhausts can be used to control the speed of the carriage motion.

In any system where high-precision shafts are used, extra care must be taken to protect these polished shafts from welding spatter. Any molten metal falling on the shaft may adhere to or pit the surface, decreasing the service life of the rail or bearings. The same is true for valves and other fixturing or positioner components that are subjected directly to welding sparks and molten metal. Welding is a very abusive atmosphere, but measures can be taken to increase the life of system components.

The conveyors shown indicate a simple method for transporting parts to and from the robotic welding cell. A very effective way to present component parts to the robot is in kits, or "kanban" style. Enough components to make a final assembly are loaded into a bin and conveyed to the robot cell. The operator receives the components, loads them into a fixture, and pushes the "go" button. After unloading the welded assembly, it is transported on another conveyor to the next operation, say, cleanup, paint, or another welding process. To truly envision world-class performance from your welding system, you must consider how the materials will flow through the cell. We will discuss material handling in more detail later in this chapter.

You do not need to use two identical types of workstations in a robotic welding cell. Flexibility is the name of the game, and a robot would not serve its purpose if it were so limited. This layout (Figure 9-11) shows a system with one flat fixture table and one tilt/rotate positioner.

Welding fixtures should always be supplied with dowel bushings, which mate with pins affixed to the positioner or table surface. In this example, the table could have a pattern of dowel pins and bolt-down holes in order to accommodate a wide variety of fixtures. These dowel pins are necessary because, if you ever remove a fixture, you must be able to exactly relocate it to the position it was at when the program was written. Otherwise, storing the program on floppy disc for future use would be an exercise in futility. You need the ability to remove a fixture, then at any later date, replace it, call up the correct welding

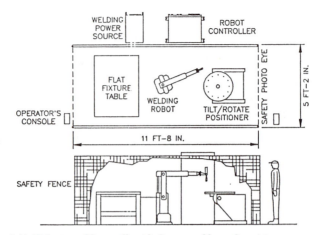

Figure 9-11. Robotic welding cell with fixture table and positioner.

program, and commence welding. This is especially important in the job-shop atmosphere where changeovers are frequent.

On the other side of the cell is a two-axis, tilt/rotate positioner. This type of positioner is quite versatile, and can be provided with both axes servo-driven and programmable, or with two simple indexing axes, or any combination in between. The disadvantage to such a positioner is that, when the platter is tilted, the weldment actually moves through space, as its center of gravity (C.G.) tilts forward. Unlike an orbital—which more or less maintains the C.G. of the assembly in a relatively confined space—the tilt/rotate positioner may not always make the best use of the robot's working range. An advantage, however, is its cost. These two-axis positioners are normally less costly than a two-axis orbital of the same weight handling capacity.

Remember our rule about loading and unloading at a second station, internal to the welding process? Rules, of course, are meant to be broken. The layout of this cell may not be ideal if only one operator is to run it. It is obvious that he would do a lot of walking back and forth between cycles. This type of arrangement might be most advantageous, however, in the following scenario. Say the tilt/rotate positioner is to handle a part with a very long cycle time of 2 hours. If the load/unload time is kept short—say within 10 minutes—

then the load/unload time is a small percentage of the total cycle time. Most of the work in this cell is done on the positioner. On those occasions when a few small, simple weldments must be done, however, then a fixture can be quickly slapped onto the table, and the robot can weld these parts during that 10-minute positioner load cycle.

Since the investment in a flat steel table is miniscule, you can afford to use it only 10% of the time, and, since the load/unload time for the positioner is a small percentage of total cycle time, then the investment in a second positioner may not be needed. Is it really possible, however, to keep load/unload time to only 10 minutes for such a large, complex weldment (one that requires a robotic cycle time of 2 hours)?

Here is how it can be done. Say the positioner is meant to weld only two types of weldments—a riding lawnmower frame and a mower deck. The two fixtures are built as normal, but, instead of connecting the base plates of the welding fixtures directly to the positioner platter, they are designed to be affixed with quick-release hydraulic clamps to a subplate. This subplate, that stays attached to the positioner platter, has a common dowel pin pattern which mates with the base plates of either welding fixture. As the robot is welding one of the assemblies, the operator is reloading the other fixture off-line, performing some tack welding, and preparing the fixture for the robot. As soon as the robot is finished welding on the lawnmower frame, the hydraulic clamps are released, and the fixture is lifted off with a hoist—leaving the subplate on the positioner. The other fixture is lifted onto the subplate, the hydraulic clamps are actuated, and the robot begins welding on the mower deck weldment. This should all take no more than 3 or 4 minutes, which adds up to only about 3% of the total cycle time. Can you afford this 3% loss of productivity? If the alternative is to purchase another $40,000 positioner, then this scenario can be quite attractive.

An alternative to the layout shown—with the two workstations opposed to each other—would be to arrange the two stations 90 to 120 degrees from each other. This would decrease the distance traveled by the operator between stations. Care must be taken, though, to maintain the operator's safety. If the two stations are placed too close together, then the operator may be in danger of being struck by the robot arm while loading parts.

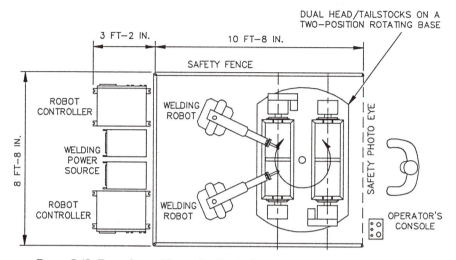

Figure 9-12. Two-robot welding cell with dual positioner.

This system configuration is an alternative to that shown in Figure 9-5. The two individual head/tailstock (HTS) positioners have been mounted on a common, rotating base, which allows the operator to concentrate on loading and unloading rather than running back and forth. The advantage can be significant, but the heavy-duty rotating base adds significant cost to the system.

These rotating bases can be either pneumatically or electrically driven, although pneumatics are normally only used for smaller loads. Much more common is the electric drive, since it is more controllable and can be more cost-effective when used to index heavier loads. This indexing motion is most often used only for station interchange but can also be programmed to stop at various angles from the robot, in order to provide more flexibility. This type of indexing base, whether electric or pneumatic, should also be equipped with a shot-pin or similar locking assembly which, when the indexing base arrives at its destination, will lock into a tight-tolerance locating bushing. This guarantees that the welding fixture will be located accurately with respect to the robot.

This cell also shows two robots working at the same positioner. This is somewhat different from the dual-robot system shown in Figure 9-9, since the robots there work at independent stations. Here, they are working on the same fixture, so the cross talk between the two robots must be well thought out and effective, in order to avoid crashes. An additional challenge when working with dual robots is to balance the program time equally between the two. Although one robot may have more linear inches of weld to do, the other may have more individual welds to make. A balance must be achieved, since you do not want your expensive robot to sit idle any longer than necessary.

An advantage to such a setup is that each robot can be welding with a different type of welding wire. For example, a single weldment may require two types of wire, say, where two components are being welded together by one of the robots with flux-cored wire, then one of the edges of the assembly is surfaced with hard-facing wire by the other robot. The two robots can complete all of the welding, with only one handling. Another common example is where flux-cored and solid wire are both required on the same weldment.

Safety is especially important when using such a large indexing base. When these big indexers start turning, nothing can stop them (except that innocuous little "E-stop" button). Although they turn slowly, the torque is quite impressive, so it is imperative that operators cannot be near it when stations are exchanging. A photo eye is shown, but many robotic equipment manufacturers use redundant photo eyes. Both photo eyes must be accounted for in order for the cycle to continue. If a photo eye is broken, then the indexing stops immediately. This also necessitates a very sturdy indexer drive train, since it takes a lot of force to stop such a large mechanism instantly. One more thing to remember: it should be a quick and easy exercise to recover from an E-stop (Emergency stop). Ask your automation vendor just what is involved in order to get going again after you have pushed the notorious red button.

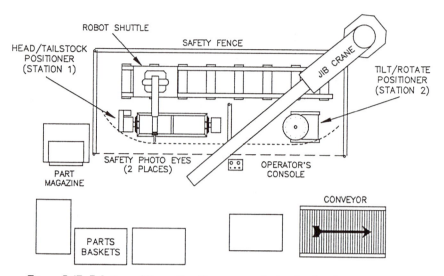

Figure 9-13. Robotic welding cell with servo track, head/tailstock, and tilt/rotate positioners.

Now we step into the world of moving the robot around as well as positioning the weldment. This cell shows the robot on a track, which provides about 12 feet of robot motion. The robot track can be either an indexing shuttle, providing one stopping position at each end of the track, or a servo track, which is fully programmable, allowing an infinite number of stopping positions.

If the only function needed is to move the robot from one welding station to the other, then a two-position shuttle track is sufficient. These are pneumatic tracks actuated by very long cylinders. Air pressure holds the robot carriage against a stop at each end for accuracy, and they can—and should—also be supplied with locking shot-pins to guarantee accurate location. If the robot carriage is out of position by a mere ¼ in., this can easily ruin the entire welding program, since all welds will miss their mark by the same amount. The advantage to indexing shuttle tracks is their relatively small cost compared to servo tracks.

Pneumatic indexing shuttles are also available with three positions—one on each end of the track and one somewhere in between. This provides added flexibility, but the middle position is typically held in place by air pressure. If you go this route, be sure that the intermediate position is locked into place with a shot-pin during welding, to guarantee the positioning of the welds.

The servo track is the most flexible method for expanding a robot's working range. The servo track is infinitely programmable, can be programmed to stop and weld in any position along the track, and all these positions are stored in memory in the same manner as the robot's positions. Servo tracks can phenomenally increase the flexibility of a robot, sometimes allowing the user to purchase less sophisticated—and therefore, less expensive—positioners. The key advantage in utilizing a servo track is not necessarily its ability to transport the robot to a number of different welding stations but to allow the robot access to the weldment from a variety of angles. This could mean the difference between the robot reaching 70% of the welds on a given weldment or reaching 85% of the welds. This will make a substantial impact on the productivity of your system.

The additional positioning flexibility provided by a servo track is especially meaningful in the layout shown, since the cell utilizes two types of positioners—a head/tailstock, and a tilt/rotate positioner. The variety of part positioning makes additional demands on the work range of the robot, so the servo track helps the robot keep pace with the more sophisticated positioning motions.

In this more complex welding cell, material handling becomes more than a mere exercise. After purchasing such a flexible welding system, poor material flow can rob you of some of that investment potential. I have shown a jib crane, since the parts being welded in this cell are quite heavy. The hoist is able to reach both positioners, as well as the finished parts conveyor. There is nothing special about jib cranes; gantry-style hoists are just as effective, depending on your individual shop requirements, the design of the welding system, and other factors. In this case, subcomponents are welded on the HTS positioner, and then these welded subassemblies are joined with other components on the tilt/rotate positioner for final welding.

A part magazine is shown to the left of the cell. This is to emphasize the fact that, unlike manual welding, the operator is now being paced by the robot. Anything that makes the operator's task easier and quicker to do will decrease his fatigue and increase the cell's productivity. Some parts can be difficult to dig out of a basket, and organizing them into a magazine can be a simple but effective way to speed things up. Parts baskets located throughout the cell can be substituted with conveyors, if it is possible for components to be conveyed directly from previous processing areas. Cellular manufacturing will lend itself much more to this concept than batch processing.

The safety equipment shown includes a safety photo eye for each of the two workstations. When the robot is welding at station 1, then the operator is free to walk through the light at station 2 and complete his work. However, if he were to break the light beam at station 1 while the robot was also at station 1, then the robot, track, and positioner would all stop immediately, shutting off the weld, and waiting for further instructions from the operator. The safety fence surrounds all three sides of the cell to prevent access to the robot by unsuspecting or uninformed personnel. Also, notice the safety fence located between the two work cells. This not only is to prevent arc glare from reaching the operator during loading and unloading, but also to prevent him from walking from station 1 into station 2 behind the photo eyes.

A common operator's console can be used for both stations. In addition to the normal complement of controls, the operator's console would provide "cycle start" buttons for both station 1 and station 2. This simplifies the cell and can eliminate some walking by the operator.

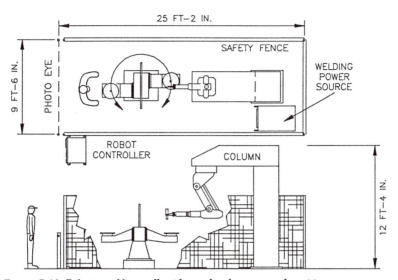

Figure 9-14. Robotic welding cell with overhead mount and positioner.

We have now learned that, by inverting a robot, its working range can be enlarged and made more effective. This drawing shows just such a scenario, with the robot mounted on a stationary column. Since the robot is stationary, the positioner chosen for this cell is a dual positioner on a rotating base. This provides a built-in safeguard by keeping the operator out of the robot's working range at all times.

The column to which the robot is mounted must necessarily be rigidly built. When the robot suddenly decelerates from a speed of 3 m/sec, this shock must be absorbed and dampened by the column, or the robot will be swaying as it starts its welding. This, of course, can lead to weld quality problems. Most such columns have an inner support system of girders and trusses welded inside in order to stiffen the structure and reduce reactions to dynamic loading.

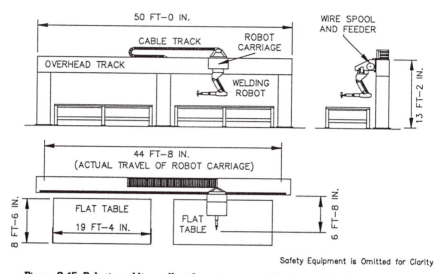

Figure 9-15. Robotic welding cell with gantry-mounted robot and flat fixture tables.

When the robot must be inverted and must move from station to station, an overhead track is the solution. The overhead track system consists of a carriage to which the robot is mounted, a long rail system on which the carriage rides, a cable track to organize and guide the cables, and upright columns to support the track.

The overhead track functions much like a floor-mounted track but must be much more rigidly built, not having a solid floor for a foundation. Not only must a robot be supported—it may weigh several hundred pounds—but the carriage also must be supported. The total load on these bearings can exceed half a ton, and all the while an acceptable repeatability and accuracy must be maintained. If the robot boasts a repeatability of ± 0.004 in. but the track carriage can only hold 0.020 in., then the robot's accuracy is meaningless. Needless to say, overhead track systems can be quite expensive, but many such systems are sold each year, so attractive returns on such an investment are available, if used in the correct application.

Ideally, the welding wire should travel along with the carriage. If a bulk spool of welding wire is placed on the floor, then the wire must be pushed, pulled, and shoved through as much as 50 feet of conduit or more. Every inch of additional conduit through which the welding wire must be pushed increases your chances of feed problems, and inconsistent wire feed will definitely harm weld quality.

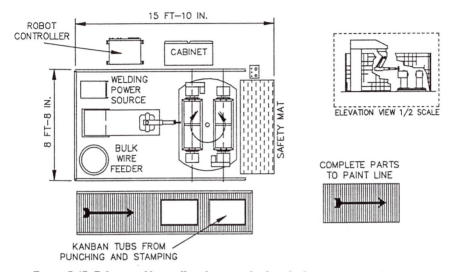

Figure 9-16. Robotic welding cell with inverted robot, dual positioners, and station exchanger.

This system is very similar to the one shown in Figure 9-14, but with a different positioner. Dual head/tailstocks are shown on a rotating base, providing safety for the operator, the ability to load and unload internally to the welding cycle, and the increased flexibility provided by inverting the robot. The bulk welding wire is located on the floor in this drawing but can also be mounted to a platform welded to the column itself. This would reduce the amount of conduit through which the wire must be pulled and would save floor space as well.

The way material flows through a robotic welding cell can vary dramatically, but the Kanban system of "kitting" parts can complement the flexibility of the robot. Kitting up parts in a tub eliminates the need for several parts baskets in the robot cell, and allows work-in-process to be reduced. Batching parts is no longer necessary, and we are one step closer to just-in-time manufacturing. Flexible fixturing may even allow several different part numbers to be welded sequentially with no changeovers. As shown, finished weldments continue along a conveyor to the paint line, the final production process.

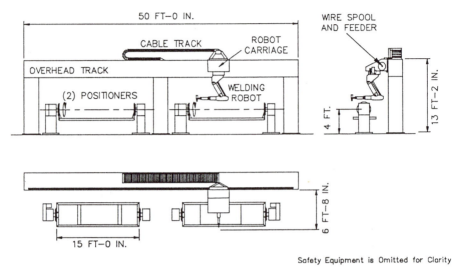

Safety Equipment is Omitted for Clarity

Figure 9-17. Robotic welding cell with gantry-mounted robot and head/tailstock positioners.

Overhead track systems can, of course, be used with positioners as easily as floor-mounted systems. In this example, parts that are much too large to be reached by a stationary robot are easily reached when the robot is mounted on a track. And the overhead mount provides even greater flexibility. In this manner, the overhead track is not only used to index the robot from station 1 to station 2, but also to move around within a single workstation, enabling it to reach a large percentage of the welds.

The distance between the positioners and the towers for the overhead track can be significant. While welding very large parts, the robot or carriage may be struck by the part or fixture as the positioner rotates them. To avoid this, the positioner must be placed further from the robot, which will cause some loss of effective working range. This factor is important in determining the configuration of the cell.

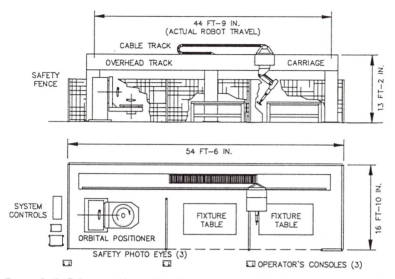

Figure 9-18. Robotic welding cell with gantry-mounted robot and various workstations.

This overhead track system is shown with three workstations—two flat tables and an orbital positioner. Each of the three stations has its own safety equipment and its own operator's console. In this case, subassemblies are welded on the two flat tables, and these subassemblies are loaded into the orbital positioner for final welding. This is a very effective way to build up a more complex part and to accomplish all of the welding in a single welding cell, perhaps with a single operator. Where several manual welding stations were required before, this single flexible system can be used. Again, by using flexible or quick-change fixturing, changeover time can be reduced, work-in-process is reduced, and each complete weld cycle yields a complete final assembly.

Conclusion

This book has taken you on a journey, perhaps one that you have never taken before, or perhaps one you were afraid to take. Whatever your situation, you purchased this book for a reason—because you want to be the best. You want to be competitive, you want to improve quality, you want to cut costs and improve your products. These are noble goals, and I want to encourage you to follow through with the information provided for you here. You are now better equipped to understand, purchase, and start up an automated welding cell than your competitors are. Do not react to what your competition is doing—set the pace. By automating your welding, you are telling your customers that you are the cutting edge manufacturer, and you have the best ability to provide what they need—the best quality at the best price.

The philosophy behind automation is a topic for other books, but it is a well-known truth that automation of any type typically means cutting costs while improving quality and service. The mystery has been removed, the industry has matured, and automated welding has much to offer today's manufacturer. What have you got to lose, other than a few minutes on the phone with a few welding vendors? More importantly, what have you got to gain? To be the best widget maker in America—or perhaps the world—is what you should expect, and the automation of your welding processes is one more rung in the ladder toward that goal of success.

Index